D1044872

THE SCIENCE OF HARRY POTTER

Also by Roger Highfield

Can Reindeer Fly? – The Physics of Christmas
Frontiers of Complexity
The Private Lives of Albert Einstein
The Arrow of Time

THE SCIENCE OF HARRY POTTER

HOW MAGIC REALLY WORKS

ROGER HIGHFIELD

headline

First Published in 2002
by HEADLINE BOOK PUBLISHING

10 9 8 7 6 5 4 3 2 1

Cataloguing in Publication Data is available
from the British Library

ISBN 0 7553 1150 7

Typeset in Perpetua by
Letterpart Limited, Reigate, Surrey

Printed and bound in Great Britain by
Clays Ltd, St Ives plc

HEADLINE BOOK PUBLISHING
A division of Hodder Headline
338 Euston Road
LONDON NW1 3BH

www.headline.co.uk
www.hodderheadline.com

Contents

To the three witches who enchant me:
Julia, Holly and Doris

Acknowledgements

All sorts of fascinating, bizarre and fantastic topics bubble up from the cauldron of complex plot lines, characterisation and mythology in J.K. Rowling's Harry Potter series. What follows is my attempt to show how many elements of her books can be found in, and explained by, modern science, from game theory and palaeontology to molecular biology and general relativity. In this way, her magic can illuminate some of the darkest corners of research, and vice versa.

Very many thanks to John Brockman and Katinka Matson, who urged me to develop this idea. Thanks are also due to Penguin Putnam for backing the project and in particular to Rick Kot for his warm encouragement and support. For the British edition, I am grateful to Ian Marshall, Jo Roberts-Miller and Christine King of Headline. I also received a great deal of useful feedback from Jan Willem Nienhuys and help from Henk ter Borg of Spectrum in the Netherlands, and comments from Piotr Amsterdamski in Poland. Gulshan Chunara, as ever, provided me with invaluable assistance in the office. Other colleagues at the *Daily* and *Sunday Telegraph,* notably David Derbyshire, Robert Matthews, Robert Uhlig and the editor, Charles Moore, have helped to indulge and aid my obsession with Harry Potter in one way or another, either directly or indirectly.

Most of all, I would like to thank a number of researchers for assisting my unusual quest to uncover the science of Harry Potter. The following witches and wizards from all corners of the scientific world answered my questions, sent me papers and provided other invaluable advice. Most of the following commented on draft sections and chapters. Any remaining howlers and errors are, of course, the result of hexes, curses and confounding spells.

Many thanks to: Manuel Aguilar, Miguel Alcubierre, Stephen Aldhouse-Green, Peter Armbruster, Qasim Aziz, Mike Baillie, Bonnie Bassler, Wolfgang Behringer, Charles Bennett, Sir Michael Berry, Michael Blake, Robert Bowles, Samuel Braunstein, Robin Briggs, John Burn, Mark Cane, Linnda Caporael, Greg Chaitin, Allan Cheyne, John Clarke, Nicky Clayton, Martin Collins, Peter Coveney, Pam Dalton, Richard Dawkins, Jelle de Boer, Jeannine Delwiche, David Deutsch, David Dolphin, Larry Donehower, Don Eigler, Thomas Eisner, Ernst Fehr, Matthew Freeman, Chris Frith, Peter Funch, Andrey Geim, Robert van Gent, Sabine Gerloff, Steven Goldstein, Sir John Gurdon, Patrick Haggard, Rosalind Harding, James Hartle, Jason Head, Keith Heikkinen, Christopher Henshilwood, Bruce Hood, Michael Huffman, Ronald Hutton, Martin Ingvar, Vincent Jansen, Thomas Johnson, Steve Jones, Cynthia Kenyon, Eric Knudsen, Ron Koczor, Beth Krizek, Janus Kulikowski, Carolyne Larrington, Andrew Leuchter, James Levine, Deane Lewis, Katheryn Linduff, Gordon Lithgow, Roland Littlewood, Bruce Livett, Elizabeth Loftus, Andrea Lynn, Peter Main, Vera Mainz, Robert Margolskee, Lord May, Adrienne Mayor, Wilfried Menghin, Emmanuel Mignot, David Moorcroft, Bill Napier, Les Noble, Sir Paul Nurse, Steve O'Shea, Jose Pardo, Alex Parfitt, Pragna Patel, Irene Pepperberg, Raj Persaud, Elizabeth Phelps, Steven Pinker, Mark Plotkin, Chris Polge, Eugene Polzik, Lawrence Principe, Martin Raff, Ismo Rakkolainen, Jef Raskin, John Rees, Martin Rees, Paddy Regan, David Reiss, John Riddle, Maria and Rick Riolo, Alberto Rojo, Igor Roninson, Mike Ross, Clinton and Jasper Rubin, Ned Ruby, Riitta Salmelin, Karl-Heinz Schmidt, Ladan Shams, Neel Shearer, Graham Shimmield, Colin

Shawyer, Dan Simons, Dana Small, Jim Smith, Gregory Snider, Noam Sobel, Ian Stewart, Jim Stone, Roger Sullivan, Paul Tacon, Karl Taube, Timothy Taylor, Kai Thilo, Kip Thorne, Andreas Trumpp, Teresa Uriarte, Julian Vincent, Fritz Vollrath, Stuart Vyse, Ian Walmsley, Diederik Wiersma, Richard Wiseman, Stephen Wolfram, Lewis Wolpert, Kielan Yarrow, Semir Zeki, Sergey Zozulya and Charles Zuker.

There are many other people to whom I owe a huge debt of gratitude. First, and foremost, my wonderful wife, Julia, for putting up with yet another book project and helping me in so many different ways to finish it on time and with my sanity intact (almost). Second, my little girl, Holly, who I have neglected. After this, her first experience of seeing her father mutate into an author, she often tells me she has to 'go upstairs to work'. My mother, Doris, and my father- and mother-in-law, Jim and Betty Brookes, also provided invaluable support, as did Claire Vernon.

Finally, a handful of people gave me help and advice that was far above and beyond the call of friendship. Graham Farmelo of the Science Museum provided crucial feedback on how to structure the book and make it more accessible. Many thanks to Matthew Freeman of the Laboratory of Molecular Biology, Cambridge, for his encouragement, good advice, creative suggestions and kind help with the Magizoology chapter. A number of friends and relatives read through drafts and came up with many useful comments, ideas and suggestions (and a few daft ones). Many thanks to Carole Gannon, Tony Manzi, Eamonn Matthews, Brian Millar (the best co-author I never had), Sharon Richmond and Martin Winn.

Roger Highfield
Greenwich
June 2002

PS Although the science in this book is real, no mythical beasts were harmed while writing it.

Any smoothly functioning technology gives the appearance of magic.

Arthur C. Clarke

INTRODUCTION

The science of magic

The most beautiful thing we can experience is the mysterious. It is the source of all true art and science.

Albert Einstein

I love the Harry Potter books, but maybe not for all the same reasons that you do. For me, every enchantment, spell, curse and other act of sorcery in J.K. Rowling's wonderful creation seems to throw down a challenge to modern science. Surely biologists would be baffled by the phut-like blast of a Skrewt? Surely brain scientists would reject the idea of a hat that can read thoughts? Fluffy, Nifflers, Hinkypunks and the other bizarre creatures that populate Harry's world seem at odds with what we know about nature. Giants, dragons and Fluffy's three slavering heads must be figments of the author's imagination. Beans of any and every imaginable flavour? Sounds very unlikely, as do flying broomsticks, the thousands of candles that hover in Hogwarts and the Weasleys' gravity-defying Ford Anglia. Surely magic of this sort can't be reconciled with the rational laws of science?

After seeking out the truth as diligently as the long pink tongue of a Puffskein searches for wizard bogeys, I think that it can. Harry's magical world can help illuminate rather than undermine science, casting a fascinating light on some of the most interesting issues that researchers struggle with today. Similarly, what we have learned from our scientific investigation in many fields can help explain many extraordinary and seemingly magical phenomena.

Finding common ground between science and magic depends, of course, on how we define magic. The simplest interpretation of magical practices is that they are bad science. But who can really say what is 'good science'? Many innovative ideas have taken years to be accepted by the scientific community. Does that mean they suddenly transform from magic into science? Does magic thrive at the frontiers of research where there is ignorance and intellectual dispute?

Another definition of magic talks of an 'art' that, by the use of spells, supposedly invokes supernatural powers to influence events. But this depends on how we define 'natural' in the first place. Are all things in the universe 'natural'? Perhaps magic has more to do with explanation than understanding, so that supernatural powers are simply those that lie beyond current human understanding. This chimes with another definition of magic, which refers to ritual activities that exert influence by access to 'an external mystical force beyond the ordinary human sphere'. But where does the boundary of that sphere fall? Somewhere near the gloomy dungeons and passages of Hogwarts? Between heaven and the stars? Presumably, this human sphere is inflating like a balloon as we build up scientific knowledge and understanding.

Perhaps, instead, the boundary between natural and supernatural lies in the human mind. After all, the perception of what constitutes 'magic' depends on the individual observer. What seems magical to one person may seem pedestrian to another. If an aborigine utters a spell before planting each and every harvest to protect crops from the elements, this ritual may have come to seem so mundane to him as to lack magic.

Equally, an urban dweller may be amazed when he catches his first glimpse of the shimmering green and red sheets of the aurora australis, an atmospheric light show in the Southern Hemisphere. Do those vast curtains in the sky count as magic?

I have been immersed in all things scientific for more than two decades, yet many events continue to strike me as profoundly magical, even though I know how to explain them. They include the first cry of my daughter, Holly; the nocturnal green tracks of fish criss-crossing a phosphorescent bay in Puerto Rico; gazing up at an unpolluted night sky from a Mars-red desert in South Africa; and, after a night working in a research nuclear reactor, bouncing a neutron off a soap bubble. These events were magical because they were beyond the familiar, commonplace and everyday.

Magic remains an elusive term. No definition has ever found universal acceptance, and attempts to separate it from 'religion' and 'superstition' on the one hand and 'science' on the other are beset with difficulties. However, as I hope to persuade you, there are some interesting connections between science and magic. They share a belief, as one mathematician put it, that what is visible is merely a superficial reality, not the underlying 'real reality'. They both have origins in a basic urge to make sense of a hostile world so that we may predict or manipulate it to our own ends. In the sense that a wizard is a wise man, wizards still exist today – and I don't just mean in the Harry Potter books and movies. Magic, like science, also gives many insights into the workings of the human brain. Both share some decidedly oddball ideas, whether jumping toadstools or quantum jumps.

The technology created by today's wizards and witches that makes aeroplanes fly and computers understand speech, and sends a faltering voice from one side of the planet to another, is sufficiently inscrutable to most people that these feats might as well be the product of sorcery. The biochemistry in a home pregnancy test, the movements of electrons around a silicon chip in a home computer and even the instructions to operate a videocassette recorder can count as magic.

A number of episodes in the Harry Potter books suggest that even its magically gifted characters acknowledge that Muggle scientists and technologists do perform a kind of magic. Take the moment that Hagrid points out a parking meter to Harry, for example. The great man is clearly puzzled by the strange device. Mr Arthur Weasley also remarks on how Muggles use considerable ingenuity to overcome their lack of magic (praise indeed from someone who works for the Ministry of Magic). He is an avid collector of plugs, batteries and anything else to do with 'eckeltricity'. For Mr Weasley, even a box of matches offers fun-filled pyrotechnics that rival any wizard display. His wife, Molly, suggests that Muggles know much more than the wizard world gives them credit for.

To the average Muggle, the revelations of modern science are often as obscure as a witchdoctor's spell, an incantation of the Dark Arts, or indeed the topics examined by a N.E.W.T (Nastily Exhausting Wizarding Test). Ironically, this lack of comprehension of what now seems a normal part of our lives, whether the workings of a jet engine or a photocopier, may even pave the way towards a belief in magic that can make seemingly impossible things possible. As the psychologist Stuart Vyse once remarked: 'Technology has taken over our lives, but science has not overtaken our minds.'

For most people, modern science, like magic, requires a leap of faith. In the days of Newton, when almost anyone could conduct practical experiments with prisms and cannonballs, science was more open to amateurs and the laws of nature were more accessible. Today scientific research has become deeply mathematical, and experiments often depend on specialist equipment ranging from billion-dollar atom smashers to gene-reading machines. Merely to analyse the results requires thousands more pounds' worth of computer equipment. Most of us have to take scientists' word for it that these calculations are correct. (No wonder, then, there is a panic when scientists admit that they are unsure of their ground, as happened when mad cow disease spread to people. No wonder that scientists take some of the blame for these mishaps – one can almost hear the cries of 'Burn the witch' and 'Burn the warlock'.)

While scientists, like wizards and witches, are the first to be interested in unusual phenomena, they would be the first to admit that much magic remains in the world. The scientific effort is relatively young and is still struggling to explain even everyday phenomena, whether turbulent patterns in a fast flowing stream, the language of the brain, or the way that the sun's energy is moved about the Earth by atmosphere and ocean to create climate.

Given this imperfect understanding of how the world works, imagine what would happen if a Muggle scientist entered Harry's world. Wearing what looks like a space suit (a biohazard suit, to protect against wizard wheezes), bristling with monitoring equipment, and no doubt armed with one of those Muggle metal wands in case of encounters with the Death Eaters, she takes a good look around at Hogwarts and its inhabitants, from the Whomping Willow to wizards and werewolves. She would probably be struck by how some of what seems magical from Harry's child-like viewpoint seems quite achievable by current Muggle technology. Think of the Chocolate Frog card where Dumbledore suddenly disappears, the mugshot of Gilderoy Lockhart which winks cheerily or the characters that inhabit Hogwarts paintings. Now conjure up the map used by Oliver Wood, captain of the Gryffindor Quidditch team, marked with arrows that wriggle like caterpillars to show game tactics. A researcher studying electronic displays would recognise that the card, photograph, paintings and map could well have been printed on electronic paper, though he may be puzzled by the interface used to update the moving images. How about the carts that rattle around in the bowels of Gringotts Bank? Once again, these would seem familiar to Muggles who are studying intelligent driverless taxis. Or spoken passwords? A routine task for voice-recognition technology.

There are many other phenomena in Harry's world that, well, seem a little less than magical to those with a scientific bent. Water-repellent coatings for glass have done away with the need for an Impervius spell. The Howler, a red envelope that contains a screaming missive, does not seem too strange in the wake of voice mail or audio files that can be sent over the internet to deliver a rambling message to someone thousands of miles away.

Muggles can now walk though walls like ghosts, thanks to a technique that creates screens consisting of a smooth sheet of fog on to which the image of bricks and mortar can be projected. The Finnish inventor Ismo Rakkolainen has proposed many potential applications of his fog walls, some of which are indeed magical. Nor does the Marauder's Map, which shows everyone's location in Hogwarts, seem so extraordinary now that Global Positioning System technology is so commonplace. Omniocular seem, well, a bit ordinary when sports fans can overdose on action replays (although they are a bargain at ten Galleons). Why does Rita Skeeter bother with a Quick-Notes Quill in this day of voice recognition technology? Why amplify a voice with a cry of 'Sonorus!' when a public address system will suffice? And why wear specs, Harry, when you can use a laser beam to carry out corrective surgery?

While such an approach may appear to be the scientific equivalent of the Disillusionment charm, which strips away magic, many features of Harry Potter's world remain magical when one takes a scientific view. Invisibility cloaks may use clever stealth technology that is only now being developed by Muggles. Broomsticks could function by switching off the tug of gravity, a feat that still seems incredible. Giants, Lobalugs,

Hinkypunks and the rest of the magical cast of characters could be the result of genetic modification, a science in its infancy, while Bott's beans could exploit new understanding of our sense of taste and smell.

But rooting around as avidly as a Niffler searching for something shiny, or as doggedly as a Pogrebin tails its victim, science can actually make the magical world of Harry Potter seem even more extraordinary by laying bare the complexity of creating a double-ended newt, the quantum jumps that colour the flash of a wand, or the mind-reading abilities of the Sorting Hat. Digging in ancient deposits, one can even find the inspiration for basilisks, giants and dragons in the bones of long-extinct creatures.

There are also magical elements in J.K. Rowling's books that are not in themselves scientific, but that chime with scientific understanding. Think of those freshly caught Cornish pixies that ran amok in Gilderoy Lockhart's Defence Against the Dark Arts classes. A thermodynamicist could well see these mischievous electric blue creatures as a manifestation of the Second Law of Thermodynamics. (Loosely interpreted, the law says that chaos rules.) Certainly the domain of magic seems to comply with this law when it comes to Dumbledore's gloomy comment that, in the end, no spell can revive the dead.

To find out just how much our world is like the magical world of Harry Potter, I contacted more than one hundred scientists from across the planet. *The Science of Harry Potter* is the result, in which I attempt to approach this subject from two distinct perspectives, as reflected in the structure of my book. Though it is not possible to split every field neatly in two, the first half is concerned more with 'how to' issues and can be read as a secret scientific study of everything that goes on at Hogwarts and the wizarding world, whether it's installing extra eyes on monsters, flying through the air on broomsticks or punishing antisocial people.

Consider those beans, nasty and nice. How does Mr Bott make them so flavourful? While our understanding of the molecular machinery of living things has led to tasty leaps in bean science, so our understanding of the genetics of development has paved the way towards manipulating a body to create a creepy Kappa, turn a croaking toad into a wriggling tadpole, or shuffle genes to make Dobby the house elf. That leads us to a round up of strange-but-real creatures and creature features, before explaining the games played in Hogwarts society. Then comes the science of the Sorting Hat, the Invisibility Cloak and other enchanted apparel.

The second half of the book focuses more on the origins of magical thinking, whether expressed in myth, legend, witchcraft, belief (both religious and magical) or monsters. It is fittingly introduced by a Foreword on the limitations of the Muggle mind. A lot of ground is covered here: the quest to turn base elements into gold and to create the elixir of life, both of which are linked to the philosopher's stone, and advances across a broad sweep of science, from the physics of atoms to the biology of all-powerful cells in the body. I assess the historical evidence for witchcraft, cauldrons and wands, and study the origins of Snape's potions. Then come the origins of mythical beasts in archaeology and natural phenomena, and the astonishing ability of the human brain to create monsters and demons. Why do we believe in magic, witchcraft and wizardry? We will see that we are, to some extent, programmed to believe and that magic will never be made extinct by the advance of science. There are very real limits to scientific prediction. There are some decidedly odd things lurking in the mathematical foundations of science. And at the ever-expanding horizon of scientific knowledge, new questions, puzzles and mysteries will continue to emerge as surely as existing mysteries are solved.

Let's get off to a flying start and reveal what the somewhat magical-sounding fields of antigravity, wormholes and quantum teleportation have to say about wizard transport, whether by broomstick, Floo powder or that muddy old footwear that turns out to be a Portkey to another place, time or dimension. As the great Albus Dumbledore would say, tuck in.

PART I

CHAPTER 1

Broomsticks, time travel and splinching

The purest mobile form, the cosmic one . . . is only created through the liquidation of gravity.

Paul Klee

'The Bludgers are up!' yells the commentator. In the airborne stadium with golden goalposts, two teams of seven players zoom around on broomsticks, swooping and weaving as they dodge their opponents' missiles – Bludgers – while trying to score with the red Quaffle. The game of Quidditch enthrals the broomstick-riding Harry, who tries to catch the Golden Snitch and win the game for Gryffindor House.

The wizarding world's favourite form of transport, the broomstick, is one of its worst-kept secrets, for by now every Muggle knows that witches and wizards use them to get about. Even now, scientists and engineers are trying to figure out how they do so. The most prized of racing broomsticks, the Nimbus 2002 and the Firebolt, probably use extremely advanced technology to defy the tug of the Earth's gravity, a technology that has massive commercial and scientific implications. Researchers from NASA would sell their grandmothers to obtain Harry's broomstick, not to mention Hover Charms, Mr Weasley's enchanted turquoise Ford Anglia, the flying motorbike that Hagrid borrowed from Sirius Black, or the candles that hover in the Great Hall of Hogwarts, all of which suggest that witches and wizards must know how to turn gravity on and off at will.

Exotic materials that can produce antigravity could also pave the way to wormholes, hypothetical shortcuts between two widely separated points in spacetime. You could, for example, step into one end of a wormhole and emerge from the other a million miles away, 10,000 years in the past. There are several episodes in the Harry Potter books where wizards travelled through a shortcut to Platform Nine and 3/4, or to visit the Diagon Alley wizard shopping arcade. Maybe they made these quick trips by wriggling through wormholes?

Enchanted travel opportunities do not end there. Harry used Floo powder to flit about. Other objects and people can appear out of thin air, whether the Knight bus, the food that fills plates at meal times, or a wizard clutching an old boot. Such remarkable materialisations could be due to exotic technology similar to that used in *Star Trek* to beam crew members of the *Enterprise* down to the surface of endless alien planets. Today, the possibility of such an extraordinary feats taking place can be glimpsed when properties of atoms have been shuffled around the laboratory by practitioners of a leading edge field called quantum teleportation.

The quest to fly with broomsticks

It is a dream as old as humanity: to step out into thin air and fly like a bird, to cast off the bonds of gravity, to soar free, zooming through the clouds with the wind rustling past our outstretched and rapidly flapping arms. Why, then, *can't* we fly? The short answer is that we are not birds. The longer one is that the human body is unable to deliver the right combination of thrust and lift. The longest answer I intend to give is that we lack feathers to help generate lift and propulsion, efficient lung design, large enough hearts, hollow bones to reduce our weight, and adequate muscle power to generate a sufficient flap.

While we cannot fly unaided, a broomstick is not as preposterous a form of transport as it sounds. Even NASA has pronounced on broomstick propulsion: a considered overview of the various technologies on offer has been put together by Mark Millis, who has the impressive title of Project Manager for the Breakthrough Propulsion Physics Project at the NASA Glen Research Center in Cleveland, Ohio.

Millis began with the oldest technology, a balloon-assisted broomstick. This does not seem like a particularly promising contender for Harry's wooden steed. First, a blimp-like construction would seem unlikely to achieve the Firebolt's quoted performance of zero to 150 mph in ten seconds. (That's fast, although a fraction of the performance of a 6,000-horsepower dragster, which can cover a quarter-mile from a standing start in less than five seconds to reach 320-plus mph.) Millis also points out that balloon-based vehicles would make easy targets for Bludgers.

How about an aeroplane-style broom? Intriguingly, this suggestion is more magical than it may at first seem. A century after the Wright brothers made their first flight, Jef Raskin, a former professor at the University of California at San Diego and inventor of the Macintosh computer, says that the usual popular textbook explanations for what keeps aircraft aloft are wrong.

Aircraft fly because air travels faster over the top surface of each wing than underneath. A theory by Dutch-born Daniel Bernoulli established that this speed difference produces a drop in air pressure over the top of the wing, which generates lift. (You can demonstrate this effect at home by blowing between two five-pound notes.) But there is a problem, says Raskin. 'The naive explanation attributes the lift to the difference in length between the curved top of a wing and the flat bottom of the wing. If this were true, planes could not fly upside-down, for then the curve would be on the bottom and the flat on the top.' But planes *can* fly upside-down, and not only do some wings have the same curve on top and bottom but even flat-winged paper aeroplanes can take to the skies.

The key question remains: how do wings generate lift? Robert Bowles of University College London, a mathematician with expertise in aerodynamics, agrees with Raskin that lift occurs when the flow of air around a wing is turned downwards. When flow is deflected in one direction, lift is generated in the opposite direction, according to Newton's Third Law of Motion. However, for a wing, it is crucial to understand that the downward flow depends on air being both deflected by the underside of the wing and bent by the topside.

The latter is trickier to visualise. Because air is slightly viscous it tends to stick to the top of the wing, particularly near the tips, and can generate whirling masses of air called vortices. You can see this effect by adding a dash of milk to black coffee and moving a spoon through it, revealing how movement through such a 'sticky' fluid generates a coffee vortex. As vortices are shed by the top surface of a wing, the flow turns downwards to generate an upward force on the wing.

With the right equipment, you could detect a force on your spoon as you move it through the coffee, says Bowles. This force – the same as the one that keeps a wing aloft – depends on the angle of attack, the shape of the spoon, and speed. Mathematical models, confirmed by generations of schoolchildren, show that even flat wings can fly if they have an angle of attack to deflect air downwards.

Although this 'airfoil theory' is now standard in books on

mathematical fluid mechanics, some mysteries of flight remain. How to capture the essence of turbulence (when air flow is disorderly) in a computer or a clever mathematical formula has in no way been mastered by even the best Muggle scientists. Turbulence is generated to some degree by all forms of flight through air. Next time you board an aircraft, just remember that Muggles still don't quite understand the magic of air travel.

Wings mark a conventional solution to the broomstick problem and one that would be a good way to build up frequent-flyer miles, though it may be easy to lose your luggage, remarks Millis, a not entirely serious answer. Apart from a mention of the Slytherin team whizzing through the air like jump jets, however, the many references to swooping and soaring on brooms contain no suggestion of wings, engines or any such equipment. Harry must sit on exotic technology.

How about a rocket-assisted broom? This is an entirely feasible solution but a stick thus outfitted could be tricky to steer and, given the long robes that wizards wear, something of a fire hazard. Which brings us to the antigravity and warp-drive brooms, a more promising approach, and a technology in which NASA seems to be very interested. Although it does not use the terms 'antigravity' or 'warp drive', Millis acknowledges that NASA is investigating related research at the frontiers of physics.

The quest for antigravity

Conventional attempts to fly have relied on generating another force to counter the tug of gravity but, so far, no one has ever found any way of 'shielding' matter from its effects. That, of course, has not stopped people from trying to turn off the most familiar force in the Muggles' universe.

One can imagine the excitement caused in 1992 when the Russian researcher Evgeny Podkletnov announced to the world in a paper in the obscure journal *Physica C* that he had shielded an area of space from gravity. The apparatus that accomplished this consisted of a cooled and magnetically suspended ring of superconducting ceramic material disc 145 millimetres in diameter and 6 millimetres thick. Podkletnov applied an alternating electric current to coils surrounding the disc to make it rotate and found that this setup reduced the weight of any object placed over it by up to 2 per cent. He observed the antigravity effect with a wide range of materials, ranging from ceramics to wood. The faster the rotations, the greater the reduction in gravity's force.

With Petri Vuorinen of Tampere University, Finland, Podkletnov submitted a second paper in 1996 to *Journal of Physics-D*. This time, however, the paper's description of his additional experiments was picked up by the media and he seems to have been accused of sorcery by his peers. Tampere University – whose Institute of Material Science was at the centre of the controversy generated by the announcement – said that it no longer had links with Podkletnov, and refused to comment on whether the antigravity device functioned or not. Vuorinen denied being involved in the project, the paper was not published and the work was dismissed as fantasy.

One of the hallmarks of real science is the way that, even if great scientists like Newton and Einstein had never lived, others would eventually have made their discoveries. In the case of antigravity, another scientist, Ning Li, had independently been researching gravity modification at the University of Alabama in Huntsville, and had studied the possibility that superconductors might generate unexpected gravitational effects, as predicted by Einstein's theory of gravity (general relativity). In the mid 1990s she, too, seemed to be getting somewhere – fast-spinning charged atoms in the superconductor produced a measurable change in a gravitational field – but then she dropped out of sight. Were wizards trying to silence these meddlesome Muggles?

Inspired by Podkletnov's paper in *Physica C*, a number of scientific institutions decided to take a closer look. Ron Koczor and his colleagues at NASA's Marshall Space Flight Center in Huntsville, Alabama, had taken an earlier interest in Li's work but could not determine how best to test her ideas with experiments. Podkletnov's approach seemed to be a simpler way to do the same thing. But the first attempts to reproduce his gravity-defying experiments proved futile, according to a 1997 paper by Koczor's team.

At the time of writing, Koczor was awaiting delivery of a replica of Podkletnov's apparatus, which NASA had commissioned the company Superconductive Components of Columbus, Ohio, to build with a grant of $600,000. Aware of the sceptics, of which there are very many, Koczor stresses that it is important to keep an open mind until he has a chance to test the device. (He adds: 'Please don't call it an antigravity machine. You don't know the level of heartburn and pain that would cause me.')

Other commercial organisations have stated that, though they doubt the effect is real, the implications of this research are too huge to ignore. If a souped-up version of this apparatus could be fitted on a spacecraft, rocket propulsion would be history: a nudge is all that would be required for lift-off. The same, of course, would go for a broomstick: one prod, and your toes would soon be skimming the ground.

Magnets, the Levitron and levitating frogs

One striking example of gravity defiance is found in an enchanting toy called the Levitron, which consists of a magnet, in the form of a spinning top, that can hover an inch or three above a repelling magnetic base. At first sight, the Levitron seems truly magical. We have known that such a device should not function since 1842, when Samuel Earnshaw of St John's College, Cambridge, published a paper showing that levitation should be impossible using stationary magnets. The American inventor Roy Harrigan was assured as much by Muggle wizards, who warned him that he was wasting his time by trying to defy Earnshaw's theorem. Fortunately he ignored them and, like a true magician, pulled the Levitron out of his hat two decades ago. The toy was then developed by Bill Hones of the company Fascination Inc. As if to underline its magical ability, the toy's patents referred to how its stability depended on the way it spins like a top but missed one important scientific point. Although this explanation actually violates Earnshaw's theorem, the Levitron's ability to hover patently does not.

A convincing scientific account of how it works had to wait until a 1996 study by Sir Michael Berry. Working at the University of Bristol, Berry is one of the wizards of quantum mechanics, the most revolutionary scientific theory of the past century, which was developed by European physicists who realised that the previous theories of physics did not hold true for subatomic particles, such as electrons.

The 'antigravity' force that repels the Levitron's top from the base is magnetism. Think of the base magnet with its north pole pointing up, and the top as a magnet with its north pole pointing down. As anyone who has played around with magnets knows, there is repulsion between two north poles, which balances the downward tug of gravity – and makes the Levitron float.

However, in order for the toy to function, the top has to spin; otherwise, the magnetic force would flip it over. Then its south pole would point downwards, and the force from the base would be attractive – that is, in the same direction as gravity – and the top would fall. The tricky part for Berry was explaining how a slight horizontal or vertical movement of the Levitron produces a force pushing the top back towards the point about which it gently bobs and weaves. It is precisely because it wobbles (technically speaking, the top 'precesses') that it does not violate Earnshaw's theorem. In recent decades, one of the building blocks of atoms, the neutron, has been trapped using a similar effect, so the theory has implications far beyond magnetic toys. However, there are no references in Harry Potter to spinning broomsticks, so there must be another way to overcome Earnshaw.

Enter the curious case of the levitating frogs which, once again, blurs the distinctions between science and magic. The feat was carried out by Andrey Geim while at the Nijmegen High Field Magnet Laboratory in Holland, working with Peter Main and Humberto Carmona. The team suspended a frog in mid-air without use of mirrors, strings, sleight of hand or any other trickery. They defeated the force of gravity with a balancing force of magnetism rather than attempting to turn gravity off at its source. 'This is, in fact, as close as we can – probably ever – approach the science-fiction antigravity machine,' they say.

The floating frog is impressive proof of something that most of us do not realise: that it's not just metals that respond to magnetic fields. The team has repeated this uplifting feat with grasshoppers, fish, mice and plants. In fact, it is possible to levitate magnetically every living creature due to an omnipresent form of magnetism called diamagnetism.

This kind of levitation is not ruled out by Earnshaw's theorem, unlike other types of magnetism: paramagnetism and ferromagnetism. Diamagnetism is a quantum phenomenon that cannot be explained by the classical physics of Earnshaw, and it turns out that everything from wood, grapes and water to pizza, frogs and even humans can be lifted by a magnet – providing it is strong enough.

All everyday materials are made of atoms, two hundred thousand million million of which would fit on the full stop at the end of this sentence. And all kinds of magnetism rest ultimately on the behaviour of electrons in atoms. Traditionally, atoms have been described as miniature solar systems, in which negatively charged electrons fly around the positively charged atomic nucleus like tiny spinning planets. (Today, we think of electrons as a negatively charged mist, rather than as discrete particles.)

Because electrons are electrically charged, their motion can generate magnetic fields. In this way, they turn into magnets that can themselves be affected by magnetic fields. When a magnetic field of sufficient intensity distorts the electron orbits in the frog's atoms, they generate a tiny net electric current, which, like an electromagnet, generates an opposing magnetic field. Like opposing magnets, the repulsive force pushes the fields apart.

'There is a sense in which the Levitron and the frog are the same, since diamagnetism is microscopically the result of tiny rotating magnets – little versions of the Levitron's spinning top,' says Michael Berry, who has worked with Andrey Geim, now in the University of Manchester, to extend his theory to show how levitating amphibians also defy Earnshaw.

When the little frog underwent this form of levitation it looked comfortable inside the magnet and, afterwards, happily rejoined its fellow frogs in the laboratory's biology department. Geim's research team has been exposed to high fields, as has one of his American colleagues who spent several hours inside a magnet (reclining, not levitating), and none of them has suffered any ill-effects. Geim has even levitated a hamster, called Tisha, who went on to live to a healthy old age of three (remarkably, they co-authored a levitation paper in the journal *Physica B* by A. K. Geim and H.A.M.S ter Tisha). Because there are no signs that these strong static fields have any health effects, Harry could easily be carried aloft this way. All it requires is a big enough magnet. The frog was lifted two metres up a cylinder by a magnetic field 100,000 times stronger than the Earth's natural magnetic field and between 10 and 100 times stronger than refrigerator magnets.

The natural pose when riding a broomstick – leaning forward, so the body is more horizontal than vertical – is in fact the best posture for magnetic levitation. The catch is that you would not really need a broomstick at all to exploit diamagnetism, and you would have to be inside a vast magnet several metres across that could generate many times the field currently used by the magnetic resonance imaging scanners that are commonly used in medicine.

'I would enthusiastically volunteer to be the first levitatee,' says Michael Berry. 'To be levitated in this way could be an interesting experience . . . more like the weightlessness experienced by astronauts in space. But there is a difference: the diamagnetism of the body is not quite uniform – tissues, bone, blood and so on have different magnetic properties – so we would feel slight pullings and pushes over the body. If the magnetic force on flesh is greater than that on bone, it would be as though we were held up by our flesh, with our bones hanging down – a bizarre reversal of the usual situation, and possibly the basis for an (expensive) type of face-lift.'

Intriguingly, one of the Potter books contains a fleeting reference to how there is too much magic in the air around Hogwarts for electronics to work – a tantalising hint that the school is bathed in an electromagnetic field powerful enough not only to disrupt sensitive microchips but also to lift a person into the sky. However, so strong a field would also exert an extraordinary tug on anything ferromagnetic, such as iron, cobalt and nickel, making its presence obvious and something of a nuisance to the inhabitants. Kennilworthy Whisp also points out that no spell yet devised allows wizards to fly unaided in human form. We may have to look elsewhere to find the secret of the Firebolt.

Cosmic antigravity

In other fields of physics, antigravity is beginning to be taken seriously. Strangely enough, Einstein himself formulated the idea, then abandoned it. Today, however, there are hints that his first instincts were correct. There is growing evidence from our studies of the heavens to suggest that mysterious 'dark energy' may be shoving huge collections of stars – galaxies – away from one another, which sounds as ominous as anything out of the pages of the Harry Potter books.

Einstein's 1915 theory of gravity came about after he realised that a falling person would not feel his own weight – until he hit the ground, that is. The force of acceleration matches that of gravity so precisely that the faller's sensation of weight is cancelled out. Thus gravity and acceleration are equivalent. Einstein recognised that his 1905 special theory of relativity had to be generalised so that it could describe varying accelerations found in real-life situations, when gravitational fields are not uniform.

General relativity replaced the previous way of describing gravity devised by Isaac Newton. Einstein did not view gravity as a force, as Newton did, but the curvature of a four-dimensional mixture of space and time called spacetime. While special relativity deals only with flat spacetime, general relativity deals with spacetime that has been warped by gravity. For example, spacetime around the Earth is warped into a shape like the inside of a bell, so that falling objects are toppling into the bell and orbiting satellites and spacemen are rolling around within it.

Einstein's theory suggested the universe was dynamic: it would either expand, then collapse under the relentless pull of gravity, or it would continue to expand for ever. However, like many scientists of his time, he assumed the universe was not contracting or expanding, but unchanging. To make his theory predict a static universe, he added a fudge factor, something he called the 'cosmological constant', representing antigravity, though he had no idea if it was real. Einstein said later that his introduction of a cosmological constant was the biggest blunder of his career. He had missed the chance to predict what the American astronomer Edwin Hubble discovered in 1929: the universe is expanding. However, recent experiments suggest that Einstein's original 'repulsive suggestion' of antigravity was on the right track and that the expansion of the universe recently began speeding up, as if something were pushing it: the antigravitational force of dark energy seems to be loosening gravity's grip. The source of this repulsive gravity is unknown. It may be something entirely new and these first observations may mark the start of efforts to grapple with a puzzle as mysterious as the pitch-black monolith that stars in Stanley Kubrick's masterpiece *2001: A Space Odyssey*.

Over huge distances this force becomes something to reckon with, and is strong enough to dent the effects of gravity. However, there is scepticism that it will be possible to harness this force to lift a broomstick. 'The only more or less accepted fact about antigravity (and there is not universal agreement) is the evidence for the acceleration of the expansion of the universe, which points to a "cosmological constant" or "energy field" (normally called quintessence) which produces antigravity,' says Miguel Alcubierre of the Universidad Nacional Autónoma de México. 'The effects, however, are extremely small.'

Wriggling through wormholes

If we could find materials that had enormous non-gravitational tension, or pull, another form of transport might become possible. (Although a rubber band shows this kind of behaviour, as do electric fields, their pulls are just not strong enough.) In that case we might be able to create wormholes, which are the cosmic equivalent of their counterpart in an apple. But rather than providing a route between core and peel, cosmic wormholes provide shortcuts between two distinct points in spacetime, separated by five miles or five million miles, five years or five million years. These could explain some of Harry's more spectacular journeys, such as the one he took through Tom Riddle's diary. The description of his falling into one page, June the thirteenth, sounds very much as though he was passing through a wormhole portal. Wormholes may likewise snake to the sorcery shops of Diagon Alley or from King's Cross Station to Platform Nine and 3/4.

To understand wormholes, we must turn to Einstein's theory of gravity, as warped spacetime. Then we have to look at what happens as a result of the extreme spacetime distortions caused by the biggest gravitational tug of all, that occurring around black holes, where gravity is so intense than even light cannot escape. While working with Nathan Rosen in Princeton in the 1930s, Einstein had discovered that the equations of relativity show that a black hole forms a bridge between two places/times (regions of spacetime). Such an 'Einstein–Rosen bridge' – which we now call a wormhole – could lead to the possibility of movement through vast distances across the universe, or even time travel.

At first sight it seems unlikely that wizards use black holes for transportation. For one thing, the wormhole itself does not exist for long enough. In effect, gravity quickly slams this portal shut. This proved to be a headache when the late astronomer Carl Sagan decided to write a science fiction novel, *Contact*. Sagan wanted to fix this problem so that his heroine could travel from Earth to a point near the star Vega. In 1985 he approached Kip Thorne at Caltech for help, who in turn enlisted the aid of his students.

They tried to work out what kinds of matter and energy would be needed for the feat of interstellar travel. In 1987 they reported that for a wormhole to be held open, its throat would have to be threaded by some form of exotic matter, or some form of field, that would exert negative pressure and have antigravity associated with it.

Thorne recently said researchers were still studying whether it was possible to get enough exotic matter in the mouth of a wormhole to maintain its gape. But the bottom line, at present at least, is that it looks as though it will be a difficult feat. 'I regarded it as fairly negative a few years ago and it has become more negative,' declared Thorne. However, he admits that his scepticism that wormholes will ever be established to be feasible for travel is not the final word on the matter.

That is just as well for Harry Potter. His experiences do seem to mirror those described in *Contact*. Sagan describes travelling through a wormhole as racing down a long dark tunnel. After a pinch of Floo powder was thrown into the burning flames of a fireplace, Harry described how he felt as if he were being sucked down a giant plug hole. As he spun around, blurred fireplaces flashed past him while bacon sandwiches churned within him. Similarly, Sagan referred to the texture of the tunnel walls as they flashed past, from which it was possible to sense the incredible speed, one at which even a collision with a sparrow would produce a devastating explosion.

Hermione's Time Turner

Physicists who have studied Einstein's theory are sceptical about the possibility of time travel but can't dismiss it. However, as Stephen Hawking once confessed, it is tricky to speculate openly about methods to whiz about in time. One risks either an outcry at the waste of public money or a demand that the research be classified for military purposes. 'There are only a few of us foolhardy enough to work on a subject that is so politically incorrect in physics circles. We disguise the fact by using technical terms – such as "closed timelike curves" – that are code for time travel.'

Kip Thorne, for example, believes that asking questions about time travel will help scientists better understand time itself, which remains mysterious. Because space and time are treated on an equal footing by Einstein's equations, a wormhole that takes a shortcut through spacetime might just as well link two different times as two different places. Einstein's theories contain nothing to forbid time travel.

But that is not the same as saying that such a trip is possible. Einstein's theories are not the whole story. They fail to include quantum mechanics, and it is quantum mechanics that is likely to make time travel impossible. Hawking, for example, claims that while wormholes might be created, they cannot be used for time travel; even with exotic matter stabilising the wormhole against its own instabilities, he argues, inserting a particle into it will destabilise it quickly enough to prevent its use. He has expounded this in what he calls his Chronology Protection Conjecture.

With Michael Cassidy, Hawking investigated what are called rotating Einstein universes, which admit time loops – time travel – and found that the probability of their having sufficient warping for a time machine to function is almost zero. This makes the universe safe for historians. 'Moreover,' he says, 'we have no reliable evidence of visitors from the future. (I'm discounting the conspiracy theory that UFOs are from the future and that the government knows and is covering it up. Its record of cover-ups is not that good.)'

However, Muggles who defend the notion of time travel point out that it is not possible to use a time machine to go back in time to before the time machine was built. You can voyage to the future, and come back to where you started, but no further. This may explain why no time travellers from our future have yet visited us.

Another objection to time travel is that it will create paradoxes. When Hermione Granger was given the Time Turner, which looks like an hourglass, so she could cope with an impossible timetable, she was warned by Professor McGonagall about the grim consequences of meddling with history, particularly the risk of killing a past or future self by mistake. This is a reference to the grandfather paradox, where you would travel back into the past and kill your grandfather, so that your mother, and therefore you yourself, were never born. In which case, you could not have gone back in time to kill your grandfather . . . and so on. No wonder the Ministry of Magic takes a close interest in the use of Time Turners. No wonder changing history is forbidden by wizard law.

David Deutsch of Oxford University counters that there is no grandfather paradox because time travel shifts between different branches of reality. He bases his claim on the so-called 'many-worlds' formulation of quantum theory. This was first glimpsed by the great quantum pioneer, Erwin Schrödinger, but actually published in 1957 by Hugh Everett III. Everett was wrestling with the problem of what actually happens when an observation is made of something of interest – such as an electron, an atom or the movement of Harry Potter on a broomstick – with the intention of measuring its position or its speed. In the traditional brand of quantum mechanics, a mathematical object called a wave function, which contains all possible outcomes of a measurement experiment, 'collapses' to give a single real measurement. Everett came up with a more audacious interpretation: the universe is constantly and infinitely splitting, so that no collapse takes place. Every possible outcome of an experimental measurement occurs, each one in a parallel universe. As one wag put it, this theory is cheap on assumptions but expensive on universes.

Parallel universes have been much exploited by science fiction writers, who one could argue were the first to propose the idea. In the classic 1937 sci-fi novel *Star Maker*, Olaf

Stapledon wrote: 'Whenever a creature was faced with several possible courses of action, it took them all, thereby creating many . . . distinct histories of the cosmos. Since in every evolutionary sequence of the cosmos there were many creatures and each was constantly faced with many possible courses, and the combinations of all their courses were innumerable, an infinity of distinct universes exfoliated from every moment of every temporal sequence.'

If one accepts Everett's interpretation, our universe is embedded in an infinitely larger and more complex structure called the multiverse, which as a good approximation can be regarded as an ever-multiplying mass of parallel universes. Every time there is an event at the quantum level – a radioactive atom decaying, for example, or a particle of light impinging on your retina – the universe is supposed to 'split' or differentiate into different universes.

Accordingly, when Hermione travels backwards or forwards in time, she may actually be arriving in a different parallel universe. 'We do not create a new reality. We just go to an existing reality and make things happen there,' explains David Deutsch. 'And they start happening when we arrive, not when we do the "paradoxical" thing.' In this way, the 'many worlds' interpretation of quantum mechanics allows a time traveller to alter the past without producing problems such as the notorious grandfather paradox.

Portkeys and how to Apparate

Another way that characters whiz around the wizard world is by conducting the tricky feat of Disapparating and Apparating: disappearing from one place and reappearing almost instantly in another. To do so, they must first pass a test to obtain a licence from the Department of Magician Transportation, reflecting a prevalent fear among wizards that if one does not

Apparate properly, he could end up splinched – leaving half of himself behind.

For those who can't Apparate (it can't be done within the grounds of Hogwarts), or who fear splinching, there are Portkeys. Harry described how he used one to travel in a howl of wind and a swirl of colour. These apparently innocuous objects – such as an old newspaper, a drink can, rubber tyre or punctured football – can transport a wizard from one spot to another at a prearranged time. (Portkeys don't always have to be mundane objects – the Triwizard Cup turned out to be one too.)

What if Portkeys and the ability to Apparate rely on a transporter of the kind used by Captain Kirk to beam down from the starship *Enterprise* to an alien planet? This line of thought, once dismissed as incredible, began to be taken seriously in 1993, when six scientists showed that quantum teleportation was possible in principle, and still more seriously a few years later when others demonstrated it in practice, in several laboratories around the world.

Teleportation has an impeccable pedigree in science, resting in part on an intellectual dust-up between two great friends and scientific giants, Albert Einstein and Niels Bohr, the Danish father of atomic physics. Their argument was triggered by quantum theory, which Einstein disliked because of its almost magical features. Indeed, if the following discussion seems baffling, that is typical for a theory which, as I will discuss in the final chapter, is outstandingly good at giving correct answers and outstandingly bad at providing a 'common sense' picture of the world.

In 1935, Einstein outlined one such perplexing feature in a thought experiment with his colleagues, Boris Podolsky and Nathan Rosen. They first noted that quantum theory applied not only to single atoms but to molecules made of many atoms as well. So, for example, a molecule in Harry's body containing two atoms could be described by a single mathematical expression, called a wave function. Einstein realised that if you then separated these two atoms, they would still be described by the same wave function. In the jargon, they were 'entangled'.

This has a strange consequence: the properties of a particle are only defined at the moment of measurement, so that as soon as you measure some quality of one entangled atom, the state of the other atom would instantaneously have altered. Measuring one entangled partner would thus define the properties of the other, even if it is at the other end of the universe. This 'action at a distance' would apparently violate Einstein's theory of relativity, which states that nothing, not even information, can travel faster than light.

To investigate this paradox was beyond the technology of the 1930s. But four decades later, studies revealed that both common sense and Einstein were wrong. Action at a distance, which was derided as 'spooky' by Einstein, does indeed occur, and it is this peculiar feature of quantum theory that teleportation exploits.

Entanglement enables anyone who wants to carry out a teleportation to sidestep one key problem, which concerned whether you could know precisely what you were teleporting. To send a complete instruction set on how to remove Harry and rebuild him in Diagon Alley, for example, would require knowledge of details (the 'quantum state' specified by the wave function) of all of Harry's constituent atoms such as their energy and position. Quantum theory would seem to say that any attempt to do this would scramble the information and prevent teleportation as a consequence of the Uncertainty Principle, which states that the more accurately you can determine the position of a particle, the less accurately you can know its speed at the same time. (The mystery of quantum mechanics is that it does not even allow you to measure the wave function of a single particle, so without a lot of identical copies you can never find out enough about it to send complete instructions.) This would seem to defeat any attempt to specify every last detail of every last atom in Harry's body.

A decade ago, Charles Bennett of IBM and others theorised that entanglement can solve this problem. They realised that the pairs of entangled atoms in effect established a 'quantum phoneline' that could 'teleport' the details (quantum state) of one particle to another an arbitrary distance away, without having to know its state. This opened up the possibility that a transporter could transmit atomic data – even an entire Harry Potter.

The subject of the first teleportation experiment was a single particle of light, a photon. The team, led by Anton Zeilinger, that accomplished the feat at the University of Innsbruck did not manage to teleport the entire particle but rather its quantum state, a property called polarisation (encountered in everyday life, for example, when light waves of only one polarisation pass through sunglasses).

First, they generated an entangled pair of photons by aiming a laser beam at a certain type of crystal. During teleportation, an initial photon (which carried the polarisation state to be transferred) and one of the photons in the entangled pair were subjected to a measurement. This measurement, in turn, entangled the initial photon with the second which itself was still entangled with its original partner. As a consequence, the remaining photon in the original entangled pair – located on the other side of Zeilinger's lab – became polarised the same way as the first photon, in a kind of quantum daisy chain. In this way, a quantum state was passed from one photon to another.

If this arrangement were used to beam Harry to a fireplace, the transporter would need a Harry 'clone' (a construction set to build Harry, if you like) that is entangled with a second 'clone' in a transporter. 'They are like identical twins who don't have a hair colour yet,' said leading teleporter researcher Zeilinger, now at the University of Vienna. 'But as soon as you observe one, and he spontaneously assumes a hair colour, the other will adopt the same.' Note, however, that entanglement is not the same as the spooky action at a distance that Einstein derided: if one Harry twin then dyes his hair Weasley red, the hair of the other twin would remain its original colour. (If you think all this is hard to understand, I regret to say that the details, such as the implications of the 'non-cloning theorem', are even more gruesome, complicated and, dare I say, magical.)

This experiment marked a key advance, although to find out if teleportation had occurred, Zeilinger's team had to destroy the teleported photon, a problem that was overcome in a more elaborate teleportation trial conducted by Jeff Kimble of Caltech, along with Samuel Braunstein of the University of Wales, Bangor, and others.

Then Eugene Polzik's team at the University of Aarhus, Denmark, made further progress by entangling groups of atoms at two different locations. 'Let's say I wish to teleport a material object containing some number of atoms in a particular quantum state from Hogwarts to your home. To accomplish this task I first need to prepare atoms of this kind at Hogwarts and the same number of atoms at your home, so that these two atomic samples are in an entangled state. This is what we have achieved for the first time,' says Polzik. 'We prepare such atomic samples by sending a special beam of light from Hogwarts to your home.'

There is even a possible quantum explanation for splinching. 'Teleporting a half or rather a distorted whole is the most likely outcome of any teleportation because it is very difficult to perform teleportation perfectly,' explains Polzik.

Remarkable as this research is, scientists have some way to go before they can build a transporter. To beam Harry to Diagon Alley, it is already clear you would require two entangled blank copies. Samuel Braunstein estimates that you would also need about ten to the power of 32 bits (a 1 followed by 32 zeros) of information about his individual atoms. Is this where the Floo powder comes in? Perhaps the powder consists of quantum dots, molecule-sized crystals of silicon that offer a cunning way to package vast amounts of information. Greg Snider of the University of Notre Dame did a few calculations for me, assuming one bit of information per dot, and concluded that a cube of dots measuring 92.8 metres (around 300 feet) on a side would be required. That is an awful lot of Floo powder to throw around.

For a second opinion, I approached one of the wizards of quantum teleportation, Charles Bennett. He points out that it is theoretically possible to store one bit of data per atom. A diamond crystal – a huge three-dimensional array of carbon atoms – could therefore store information by using carbon 12 or carbon 13 isotopes at each atomic location in the lattice of atoms. Carbon 12 would correspond to 'on', say, and carbon 13 to 'off'. Assuming we were carrying out classical teleportation – that is, it is sufficient to stipulate the location of all the atoms in Harry's body, rather than all the details (quantum state) of each one of them as well – Bennett calculates that we would require a diamond about 100 times Harry's own weight. To grow this diamond and then read the information stored in it would be a formidable challenge.

Fortunately Bennett adds that, though 'utterly impracticable for the foreseeable future', there is no fundamental reason to rule out teleportation – and Apparating wizards – completely. 'While not violating any laws of nature it would require completely unforeseeable technical advances.' Given the evidence put before us by J.K. Rowling, Bennett, his colleagues at IBM and others in the Muggle world would do well to get their hands on some Floo powder.

CHAPTER 2

How to play Quidditch without leaving the ground

I soared where my hallucinations – the clouds, the lowering sky, herds of beasts, falling leaves . . . billowing streamers of steam and rivers of molten metal – were swirling along.

Gustav Schenk, after inhaling burning henbane seed

Muggles were swooping, soaring and zooming through the air long before the first transatlantic broom crossing in 1935 (by Jocunda Sykes, according to J.K. Rowling) or indeed the first powered flight in 1903 by two of their kind. Hundreds of years ago in Europe, they were wheeling across the sky sitting on broomsticks. In Central Africa, they preferred to soar into the heavens in saucer-shaped winnowing baskets. In the Middle East, they sprawled upon flying carpets.

Even today, many people achieve the feat of flying without the fuss and bother of aerodynamics, glass cockpits and turbofans. In London, some waft as high as Mary Poppins. In California, Muggles hover above the swimming pools that are scattered around the state like pieces of stolen sky. In

Scandinavia, they follow in the hoofsteps of Rudolph the red-nosed reindeer. However, like any witch or wizard who observes the International Statute of Wizarding Secrecy, they are not particularly forthcoming about how they journey skywards.

As it happens, there is no need to attempt to defeat the tug of gravity with rockets, engines or wings, let alone magically endowed broomsticks and baskets, for humans have accomplished as much since the dawn of their history by tricking the mind. For centuries, shamans, wizards and wise men have hallucinated through bloodletting, fasting and meditation. Some reiterated syllables, as in the Hindu mantras, or danced themselves into a frenzy. The Sufis of the Middle East used the monotonous repetition of music.

There were also as many natural pharmacological ways to get as high as the competitors in Quidditch. One could breathe in anaesthetic gas or ingest a range of natural substances. Some people developed a taste for bufotenine, a toxic alkaloid from the skin of toads. Coca was venerated by the Incas. The Aztecs used psychotropic mushrooms, which were so revered that they were called 'god's flesh' (*teonanacatl*). To enter the spirit world, Waiká shamans in South America sniff a potentially lethal hallucinogenic snuff made from three plants.

Barasana Indians in the northwest Amazon valley of Colombia distinguished forms of ayahuasca, 'vine of the soul' (*Banisteriopsis caapi*), on the basis of the colour of visions it produced. The Quechua Indians in Ecuador asserted that it allowed the spirit to wander. To find out if that was true, Richard Evans Schultes of Harvard University experimented with the plant and reported that, after an initial bout of giddiness, nausea, vomiting and perspiration, his vision became disturbed, with flashes of light and a blue haze, before he fell into a deep sleep, rich in dreams.

The Huichol of Nayarit in Mexico exploited the peyote cactus to commune with their god. Today, scientists believe its visionary effects, such as hallucinations, weightlessness and an altered perception of time, are caused by the ingredient mescaline, though other alkaloids contribute. In one test of mescaline in the 1930s, a volunteer described seeing an object 'like an iridescent plum pudding suspended in the sky exactly a hundred miles above earth'.

That feeling of light-headedness which gives the illusion of flying could be caused in various ways. Colin Blakemore, a vision expert at Oxford University, points out that drugs might have a direct effect on the vestibular apparatus of the inner ear. This contains sensors that indicate acceleration and help tell up from down, using so-called hair cells, which are so incredibly sensitive they can respond to movements so small that they are equivalent to the diameter of a hydrogen atom.

The hair cells are located in three semicircular canals within the ear. Their tufts are displaced when fluid in a canal moves, activating a nerve. Your brain uses this information about how your head is rotating to control the movement of your eyes. In this way, the eyes can compensate for the head rotation, which is indispensable if you are trying to focus on a moving Bludger while you race for the golden goalposts, for example. Other patches of hair cells within the ear are heaped with tiny lumps of chalk called otoliths. The weight of the chalk tells you which way is up.

The sensitivity of hair-cell sensors that detect gravity and acceleration is dependent on their cell membranes. One reason we feel dizzy when we drink too much alcohol is that alcohol diffuses into the membranes under the otoliths and the hair cells of the semicircular canals, altering the perception of gravity and inducing the typical sensation of spinning when you feel drunk, explains Kai Thilo, also of Oxford University. Similarly, certain drugs may influence the secretion of charged atoms across the membranes or the way the hair cells activate nerves. The result is a sensation of light-headedness and movement, even when there is none. When there is a conflict between what the eyes tell the brain, and what the inner ear says, vertigo and dizziness result.

In addition to interfering with the inner ear, hallucinogenic drugs may also disrupt centres of the brain that are used to detect movement, and specifically those involved with vection – the sense of self-motion. This is the sensation felt when the train next to yours begins to move and which creates the feeling of flying in virtual reality games and in flight simulators. By interfering with vection perception, drugs can induce the illusion of soaring and swooping.

Other insights into pharmacological flying have come from experiments by Janus Kulikowski at UMIST in Manchester, in which he studied the effects on monkeys of a veterinary drug, called ketamine, that is similar to the hallucinogen angel dust. Ketamine seems to desensitise that part of the vision system of the brain which processes patterns, their shapes, form and stability, and the way they move. The brain detects two general properties: features that are sustained and those that change. Normally our vision is well calibrated so that it receives and interprets signals from our eyes corresponding, more or less, to actual events. However, if a drug damps down the 'sustained' system that picks up stationary objects relative to the 'transient' system, which detects motion, the brain thinks there is movement when there is actually none.

Whatever the precise details of exactly how such drugs do so, their mind-altering methods can summon up a mystical experience to order. They can change perception, distort time and warp reality. The ancients who could not rationalise these bizarre effects focused instead on the subjective: out-of-body journeys, communing with the spirits, soaring into the heavens, and voyaging to the world of the dead.

Visionary rocks of the omphalos

While Sybill Trelawney's divination class relied on fire omens, palmistry and astrology, a mind-warping gas was central to the visions routinely dispensed at the most important shrine in ancient Greece, the Oracle at Delphi. For those seeking guidance, the oracle was the navel of the world ('omphalos'). There, pilgrims could seek guidance from Apollo's mouthpiece, the Pythia.

Numerous Pythias held sway over the Greek world for more than a millennium, stretching back from the fourth century AD, when an earthquake destroyed the oracle, to around 800BC. (Indeed, the influence of the Pythia may have even begun much earlier, around 1400BC.) The Pythia gave cryptic answers to such pressing matters as how to lift a curse, select a leader or build a new colony. Perhaps the most memorable Pythian utterance was to the ancient Greek Oedipus, who was told he was fated to kill his father and marry his mother.

The first clues to the true inspiration of the Pythia came centuries ago when the temple's priest, Plutarch (AD46–120), explained how the priestesses became 'filled with divine breath'. Each one was said to sit on a three-legged stool in a narrow space below the temple floor, clutching a laurel branch. There, she became gripped by the spirit of prophecy while inhaling vapours. Around the time of Plutarch, however, the Oracle's power began to wane because the source of these vapours was running out.

Like Plutarch, other ancient writers referred to a fissure in the bedrock, a gaseous vapour and a spring. But when French archaeologists investigated at the start of the twentieth century, and failed to find supporting evidence – notably a fissure or cavity under the temple – the notion of intoxicating vapours as the source of the Delphic revelations was dismissed.

The ruins of Apollo's temple still stand today outside the modern village of Delphi on the southern slope of Mount

Parnassus. At the site, another attempt was recently made to trace the origin of those inspirational fumes. In spring water near the Oracle, the emissions that were probably responsible for the Pythias' trance state – gases from bituminous limestone such as ethane, methane, and ethylene – were found by Jelle de Boer, Professor of Earth Science at the Wesleyan University in Middletown, Connecticut, working with archaeologist John Hale and chemist Jeff Chanton. Sweet-smelling ethylene in this mix matches Plutarch's description of the smell of the gas the Pythia inhaled.

Henry Spiller, director of the Poison Center at Kosair Children's Hospital in Louisville, Kentucky, says that ethylene was commonly used by doctors as an anaesthetic gas from the 1930s to the 1960s, and the first stages of the effects it produced resemble those descriptions of the Pythia when, on rare occasions, she inhaled too much and went into a manic 'frenzy'.

Fungal flying feats

In January 1692, a mysterious illness struck a number of girls in Salem, Massachusetts, causing seizures, trances and hallucinations. Several admitted that they had been secretly meeting Tituba, a West Indian slave who had entertained them with tales of black magic. Soon other of the town's girls began to have frightening visions that led to fits of panic and bizarre behaviour. The local doctor could find no known explanation for their torment, and they were considered bewitched.

Tituba and other townspeople were later arrested for practising witchcraft. When Tituba confessed, the hysteria grew. A special prosecuting court convened on 2 June 1692. That same day Bridget Bishop, the first to be tried, was sentenced to death, setting in motion a ten-month terror that would leave nineteen men and women hanged, one crushed to death with stones and another seventeen sentenced to prison, where they would eventually die.

Many factors contributed to the Salem witch-hunt. Some historians point out that the town was becoming prosperous (and secular) more quickly than the surrounding rural areas. Rivalries had developed, and the charges of witchcraft came first from the poorer, more religious countryside.

Doctors have speculated that the outbursts of the stricken girls resembled the symptoms of 'post-traumatic stress syndrome' which may have been linked to the drudgery of their daily life or even rebellion against abuse by stern parents. But the most interesting theory attributes the strange events in Salem to a wet winter that encouraged the growth of ergot in the local grain supply.

Ergot is a fungus blight that, if ingested, causes sensations of burning, cramps and contortions, and a feeling that ants are crawling under the skin. Blood vessels constrict, leading to the loss of fingers, toes, arms or legs. Centuries ago, the resulting black gangrene was blamed on holy fire, a punishment for the victim's sins. Ergot, *Claviceps purpurea*, is also a hallucinogen. As was the case with the Pythia, it does not require much imagination to understand how, before the rise of science, ergot-induced visions might well have been perceived as some form of magic.

Linnda Caporael at the University of California at Santa Barbara found that many people in Salem fell ill with symptoms similar to those caused by the drug LSD (lysergic acid diethylamide), a powerful hallucinogen, to which the key ingredients of ergot bear a chemical resemblance. Indeed, they probably inspired the creation of LSD in 1938 by Albert Hofmann, a Swiss chemist, while he was studying the ergot alkaloids. He described the effects of LSD in 1943 as 'an uninterrupted stream of fantastic pictures, extraordinary shapes with an intense, kaleidoscopic play of colours'. More relevant to the events in Salem, he reported how at higher doses one of his neighbours, Mrs R, turned into 'a malevolent, insidious witch'.

In Salem, fits of possession might have been induced by the mind-altering effects of the fungus. The young girls, being smaller, could have been the most affected by tainted grain. While the weather conditions of 1691were conducive to the growth of ergot, the following year there was a drought and the epidemic of bewitchings halted.

Salem was far from an isolated incident. Many of the seventeenth-century witch panics occurred in places where rye was widely cultivated, and in the wake of weather that was propitious for the growth of *Claviceps*. John Mann of Queen's University Belfast, in his book *Murder, Magic and Medicine*, points out that the Great Fear (*La Grande Peur*), a bout of civil disobedience that struck France in July 1789, featured reports of peasants who had 'lost their heads' and references by local physicians to the cause being 'bad flour'.

Broomsticks, Mokes and Santa Claus

Another mind-altering fungus that plays a role in Muggle magic and witchcraft is the red and white toadstool that often features in fairy tales and can still be seen today in children's books and videos – dancing, for example, in Walt Disney's *Fantasia*. Known as fly agaric (*Amanita muscaria*), it was also probably the preferred recreational and ritualistic mind-altering drug in parts of northern Europe, before vodka was imported from the east. Soma, an intoxicating drink used in India, may also have been brewed using the toadstool.

The hallucinogenic ingredients of fly agaric are ibotenic acid, muscarine and muscimol. Muscarine mimics the action of a messenger chemical, called acetylcholine, on nerves. Ibotenic acid and muscimol probably interfere with the action of GABA, another neurotransmitter.

During a mushroom-induced trance, a shaman would start

to twitch and sweat. He believed that his soul left his body as an animal and flew to the other world to commune with the spirits, who, the shaman hoped, would help him to deal with pressing problems, such as an outbreak of sickness in the village. With luck, after his hallucinatory flight across the skies, he would return bearing the gifts of knowledge from the gods. Some have even linked the red and white toadstool to the origins of Christmas mythology. Santa's jolly 'Ho-ho-ho' may be the euphoric laugh of someone who has eaten the mushroom and received its visionary gift.

John Mann cites one striking account of the effects of the toadstool by George Steller, who spent several years with the Koryak tribe in northeast Siberia. 'The person becomes completely intoxicated and experiences extraordinary visions. Those who cannot afford the fairly high price [of the drug] drink the urine of those who have eaten it, whereupon they become as intoxicated, if not more so.' Another account of its use described hysteria and visions, during which some 'deem a small crack to be as wide as a door, and a tub of water as deep as the sea'. This contains echoes of what Lewis Carroll described in *Through the Looking Glass*, when Alice ate one side of a mushroom to miniaturise, and even a strange creature called a Moke, one that J.K. Rowling says is able to shrink at will.

Such was the intensity of the toadstool's mind-altering effects that it seems quite plausible that anyone who ate it would be prepared to converse with any gnomes who happen to be perching on the toadstools in question, throwing open a portal to a world of elves and fairies or, even more pertinently, to flights into the heavens on a broomstick. However, this feat does not depend on the broomstick itself but what is smeared upon it. In the fourteenth century, there are reports of witch hunters discovering 'a pipe of oynment, wherewith she greased a staffe, upon which she ambled and galloped'. In the following century, witches confessed 'that on certain days or nights they anoint a staff and ride on it to the appointed place or anoint themselves under the arms and in other hairy places'.

A witch who wanted to 'fly' to a witches' sabbat, or orgiastic ceremony, would mix hallucinogenic plants with fats or oils, so they could penetrate the skin, and would then anoint a staff, stick or even broomstick with this cocktail and use it to apply the drugs to their vaginal membranes. This would allow the active ingredients to pass into the bloodstream and then the brain. The hallucinogenic drug would have distorted their perception to the point where they thought that they were actually flying.

One of the plants used for the purpose was henbane, from the nightshade family. In ancient Greece it was believed that people under its influence became prophetic. Henbane contains alkaloids, notably hyoscine, which have well-documented effects. High doses are deadly but, in small amounts, hallucinations, delirium and a lack of coordination result. In one experiment, when the smoke of henbane seed was inhaled, the result was a feeling of falling apart, of flying and hallucinations. Others reported how it boosted night vision, which would have come in handy for nocturnal flying.

In one twentieth-century experiment, a seventeenth-century recipe for a witch's salve was made from deadly nightshade, plants from the *Datura* genus (which also contain hyoscine) and henbane, and gave its subjects dreams of wild rides. The use of a 'flying ointment' made from datura by the Yaqui Indians of north Mexico also produced a feeling of soaring into the clouds.

The real Quidditch

All the skills of flying a broomstick are put to use in the magical game of Quidditch, which supposedly got its queer name from Queerditch Marsh, the location of the very first match. Now, of course, it is the premier sport of the wizard world. Many readers of Harry Potter have commented on how this mixture of Beaters, Bludgers, Chasers, Quaffles and so on seems to be an amalgam of Muggle ball games such as lacrosse, rugby and football. But where did the idea of Quidditch first originate? And have players always taken to the skies?

I suspect the answer to these questions may lie in Central America, Mesoamerica, where an extraordinary ball game – perhaps the most amazing of all time – contains intriguing echoes of Quidditch and its American variant, Quodpot. (Indeed, I was surprised that that this does not even merit a passing mention in *Quidditch Through the Ages*, marking a serious omission by Kennilworthy Whisp.)

The people who took part in *Nahualtlachtli*, which means the magic ball game, were probably involved in the very first team sport. The game was played for thousands of years and probably started in Mexico in around 1500BC, with Mesoamerica's first great civilisation, the Olmec, according to Manuel Aguilar of California State University, Los Angeles. By 1200BC it was being played in Oaxaca, the Mexican highlands and in the west of Mesoamerica, notably El Opeño, Michoacán. The location of its birthplace is no accident, for the balls were made of rubber, which originated in Mesoamerica.

In all, these ball games were played for more than 3,000 years, and although they varied from one region to another, they were one of the defining features of Mesoamerican life, until their eventual prohibition by the Spanish in the sixteenth century. (The rubber from which the balls were made was so alien to the Old World that, after the Spanish Conquest in 1521, when the conquistadors first caught sight of these 'bouncing, noisy' objects they thought them magic, the work of the devil.)

On the great Chichén Itzá court in Yucatan, Mexico, teams competed on a playing alley with stone ball game rings on the walls. The most highly rated player was the one who managed to get the ball through the rings and, much like the capture of the golden snitch in Quidditch, the feat was so difficult that accomplishing it signalled the end of the game. The Aztec form of the game allowed only buttocks or knees to make contact with the ball, while other varieties seem to rely on bats, sticks and handstones – the equivalent of the rounders bat used by the Beaters in Quidditch.

More than 1,500 ball courts have been found and many more probably lie undiscovered. They were as important – probably much more so – to the local people as Quidditch is to the pupils of Hogwarts and provided endless inspiration. Mesoamerican artists created miniature ball courts packed with players and spectators, elaborately attired figurines of ball players, and an array of athletic equipment whose beauty and symbolic meaning provided more than just physical protection.

The ball game also features in the *Popol Vuh*, the Maya story of creation. The stars are the ball-playing Hero Twins, Hunahpú (Hunter) and Xbalanqué (Jaguar Deer), who had a score to settle with the underworld. Their forefathers, Hun Hunahpú (One Hunter) and his brother Vucub Hunahpú (Seven Hunter), had been tricked into visiting the underworld to play the game. They lost and were sacrificed. But when the Hero Twins appeared in a return match with the underworld Lords, they won, retrieved the bodies of their father and uncle from the ball court, and placed them in the sky to become the sun and the moon.

The ball game was indeed a ceremonial activity that celebrated a magical battle for survival, where a human team was symbolically pitched against the gods and the awesome powers of the natural world. Each clash was seen as a struggle between the opposing forces of day and night, good and evil, and life and death, echoing how the best games of Quidditch pit sly Slytherin against noble Gryffindor.

The game was richly textured with symbolism to reflect the creation story. The court was the door to the underworld and the channel for the sun's rebirth each day. When the ball was passed through the stone ring at Chichén Itzá, the equivalent of the golden poles with hoops at the end found in Quidditch, it symbolised the moment that the earth swallowed the sun – along with night, darkness and death.

Many of the ball court sites feature sweatbaths for ritual purification. Their remains also include Hero Twin markers, which are round and often framed by a four-leafed motif called a quatrefoil, and represent the mouth of a cave, an opening or portal to another time or place. (There are hints of such portals, or something similar, in Quidditch, in which 'referees had been known to vanish and turn up months later in the Sahara desert'.)

A blow from the heavy rubber ball used in the game could also leave players hurt or dead, just as Bludgers could cause serious injury by knocking Quidditch players off their brooms. Although there was a notoriously dirty world cup game played in 1473, people rarely died while playing Quidditch, unlike the Mesoamerican ball game, which often had fatal consequences.

Jade and stone 'perforators', some decorated with a ball player, were used in ritual bloodletting, so blood could be collected and burned with incense as an offering. (The site the players most preferred to perforate was the penis.) Human sacrifice by decapitation was also a recurring theme in ball game imagery, because the players believed that the sun needed to be nourished with blood in order to defeat the dark forces of the underworld. Two panels from the ball court at El Aparicio, a site in Veracruz, show the severed necks of two players, from which spurting blood is transformed into seven serpents, symbolising the fertility and regenerative power of blood.

Stone heads worn on the player's yoke (belt) during ceremonies alluded to opponents he had defeated. At Chichén Itzá and central Mexican sites, *tzompantli* (skull racks) were placed in the plaza outside the courts to display decapitated heads. The representations of human skulls on the ball, as at Chichén Itzá, refer to an episode in the *Popol Vuh*, when the head of Hunahpú was used as a ball. However, they also contain the unpleasant suggestion that the Maya made the rubber balls lighter by placing skulls at their centres.

Removing the heart of a losing team member was another of the game's ceremonial events. One *hacha*, a decorative stone worn by players, shows a defeated player arched backwards, awaiting extraction, while on a ball-court relief one ball player is about to remove another's heart. Another piece of art contemporary with the playing of the game shows that a rope has been knotted around an intended victim's neck, suggesting that partial strangulation may have made the job easier. In the hands of a skilled priest, using a sharp blade, the operation was relatively quick.

While ball players did not fly, did Mesoamerican societies, like European witches, make use of hallucinogens? There are suggestions this was the case. Karl Taube of the University of California, Riverside, said that the evidence of hallucinogens in classic Mayan society is scarce, though 'mushroom stones' have been found in Guatemala, some up to 3,000 years old, with hemispherical caps and elaborately carved stems that could be linked to the New World's use of psylocibine, the active ingredient of their sacred mushroom (*Panaeolus campanulatus*) – *teonanacatl* – which produces LSD-like effects. (Indeed, Hofmann, the discoverer of LSD, described the 'whirlpool of form and colour' that the mushroom sucked him into.) Some modern scholars suggest Mesoamerican societies may also have indulged in hallucinogenic enema orgies, though Taube suspects that they used alcohol.

For the Aztecs, there is strong evidence supporting the use of hallucinogens, such as morning glory, peyote, and pscilocybin mushrooms. Another possibility is that they ate the rhizomes of water lilies, which are hallucinogenic. Water lilies were symbols associated with access to the underworld because mirrors, such as a reflective surface of water, were believed to be portals to other realms throughout Mesoamerican history. The lilies also open their petals at dawn and close them at dusk, slightly submerging, reflecting how the sun was thought to die daily and go to the other world.

Maria Teresa Uriarte of the Universidad Nacional Autonóma de México, Mexico City, remarks: 'The desire to enter a different reality is a common one in humans, regardless of the time or the place to which they belong. This may be why the water lily was for the Maya . . . the symbol of access to that different reality, induced by its psychotropic effects, and, at the same time, the symbol of the still water of access to the underworld. The water lily opens its petals at dawn and closes them at sunset, which reinforces its symbolic content.'

Uriarte found evidence that hallucinogenic substances were linked with the ball game. For instance, representations of a toad which is known to produce a powerful drug are found in objects related to the game (though there is some argument over whether Mesoamerican toads are as hallucinogenic as those found in the Sonoran Desert of northwestern Mexico). Images of the lilies appear in game murals of Tepantitla in Teotihuacan. These same murals feature other flowers that have hallucinogenic properties, including the datura family and morning glories. Such flowers also appear in Chichén Itzá's reliefs of the ball game, where there are suggestions that they may have been used to drug a loser before decapitation.

Manuel Aguilar adds: 'Even though there is not any direct evidence of the use of water lilies by the Maya players of the ball game, I incline myself to think that they were ingested by the players as part of the ceremonial context. There is a parallel between the hallucinogenic properties of the water lily and the ball game that operates as a portal to the underworld. Both elements put the person in ecstatic and mystical communication with the supernatural sphere.'

This extraordinary, deadly antecedent contains so many elements of Quidditch that perhaps it is no wonder that Harry feels a knot of anxiety every time he plays an important game. But Harry has the mettle to cope, for he has the leonine qualities of a member of Gryffindor house. He is brave at heart, daring and shows nerve. The way his personal characteristics were discovered reveals more intriguing connections between Hogwarts wizardry and the magic of the Muggles.

CHAPTER 3

The Invisibility Cloak, Sorting Hat and other spellbinding apparel

I shall never forget that dawn, and the strange horror of seeing that my hands had become as clouded glass, and watching them grow clearer and thinner as the day went by, until at last I could see the sickly disorder of my room through them, though I closed my transparent eyelids. My limbs became glassy, the bones and arteries faded, vanished, and the little white nerves went last.

H.G. Wells, *The Invisible Man*

How, exactly, did the Sorting Hat divide a sea of pointed black hats among the school's four houses during Hogwarts' annual student intake? When Harry put on the hat, it could read his thought patterns and from them discern his talent, goodness and courage. Initially, the hat dithered over whether to put Harry into Gryffindor or Slytherin house, but chose the former when crackles of activity in Harry's brain revealed his hatred of Slytherin.

In this day and age, when most hospitals have a body scanner of one sort or another, the abilities of the Sorting Hat do not seem so extraordinary. Is it possible, then, to observe the conscious brain at work and read its thoughts? Surprisingly, the answer is a qualified yes. Muggle scientists are making progress in the effort to understand the language used within a living brain by studying what happens, as we think, to the patterns of electrical and magnetic activity within it, or surges of blood flow to different areas of nerve tissue.

The hat is not the only example of enchanted apparel to appear behind the mullioned windows of Hogwarts, or even disappear behind them. There is also, of course, the Invisibility Cloak. (Or there isn't, depending on your point of view.) How could the cloak possibly work? And why is it that wizards do not seem to invest much time in robe upkeep? As a schoolboy, I had to wear a sixteenth-century school uniform that included a 'Housey coat' that reached down to my ankles, had umpteen brass buttons, velvet collar and cuffs. This archaic outfit rapidly became tatty in our boisterous, adolescent all-male boarding school environment, which bore some resemblance to Hogwarts. Why are there so few references to mending, darning and replacing robes at Hogwarts?

SQUIDs and the Sorting Hat

For more than a thousand years, the Sorting Hat has known the minds of the pupils at Hogwarts. Today, scientists are beginning to find answers to how it might function. They can already study the processes in a living brain, down to events lasting a thousandth of a second, with the latest generation of body scanners. They can watch memories being knitted from networks of brain cells. It does not seem fanciful to expect that one day a scanner will be able to chart what happens as a pupil's brain alters during education.

The Sorting Hat could, for example, read thoughts by measuring the rapidly changing magnetic field that results from the electrical activity generated within the conscious brain.

This mind-reading method is called magnetoencephalography, or MEG, and records magnetic signals generated when currents flow along nerve cells. First demonstrated in 1968, it offered advantages over one traditional method – electroencephalography or EEG – that uses scalp electrodes to detect crackles of electrical activity given off as brain cells work: EEG cannot accurately pinpoint the site of the currents because the signal is blurred by the effect of the intervening tissue. To a magnetic field, however, it is as though there is no intervening skull or scalp.

The catch is that these magnetic fields are so tiny that to detect them, with any confidence, it is necessary to use a Superconducting Quantum Interference Device, or SQUID, the most sensitive measuring instrument known to science. Perhaps the Sorting Hat used an array of them to read Harry's mind.

SQUIDs can turn a change in magnetic field generated by a living brain into a change in voltage, explains MEG expert Riitta Salmelin of the Helsinki University of Technology in Finland. 'Voltage is rather easy to measure accurately while magnetic flux is not, and that is why SQUIDs are so important.'

Each SQUID consists of a ring of material, called a superconductor, which sheds all resistance to electricity, interrupted by two thin layers of insulating material known as 'tunnel barriers'. These 'Josephson junctions' are named after the British physicist Brian Josephson, who won the Nobel Prize for his research on them as a graduate student in Cambridge.

Using the field of quantum mechanics, he predicted that electricity – in a superconductor in the form of pairs of electrons – can 'tunnel' through the junctions in the superconducting ring, allowing a current to flow. When an electrical current is passed through the two junctions in parallel, it turns out that the resulting voltage is exquisitely sensitive to tiny magnetic fields, even those created by electrical activity in

Harry Potter's living brain. When Harry thinks, his neurons generate an electric current which in turn sends out both magnetic and electric fields. The magnetic fields sprouting from Harry's head influence the electron pairs circulating in the SQUIDs in the Sorting Hat. Because quantum mechanics says that all the electron pairs in each SQUID act in concert (in the jargon, all the electrons are in the same 'quantum state'), they convert a tiny change in his brain's magnetic field into a detectable change in voltage with a sensitivity unmatched by any other device.

Although current Muggle technology is not able to read specific thoughts passing through our heads, according to Salmelin, scans can tell what we are thinking to some limited extent, she says, referring to various efforts to train paraplegics to produce distinctive brain patterns that can be used to move a cursor on a screen, or even an artificial limb. Another example of rudimentary mind reading arose in an experiment Salmelin conducted on an eminent brain scientist. 'He was thinking of playing the piano, writing with the left hand, and so on. He did not tell us in which order he was doing the tasks. However, as we knew the selection of tasks, we could tell from the MEG signals which ones he had been performing.'

But it is not enough to detect brain activity alone. The problem is isolating the significant activity that represents the thought 'I want to be a member of Gryffindor' from among all the other magnetic signals, a cacophony generated by other thoughts on the wing, routine brain activity and so on. Even if a clear signal could be made out from all the 'noise', it is not possible at present to map out how each individual thought relates to a distinctive pattern of cell activity.

However, one day, with sufficient computing power and better models of how the brain works, such mind reading may indeed be possible. 'The critical thing would be to learn to model the brain and the way it really "thinks" and then be able to create efficient learning programs (in super duper computers!) which can "interpret" the brain signals,' says Salmelin. Nevertheless, by the time such feats are theoretically possible, scientists may well conclude they are even further away from accomplishing them when they come to fully appreciate the brain's staggering complexity. Every great triumph of adventure in thought has invariably spawned new questions.

The Pocket Sneakoscope and Secrecy Sensor

SQUIDs or one of the many kinds of brain scanner techniques may also be at work in the Pocket Sneakoscope, which lights up, spins and whistles when it detects that someone untrustworthy is nearby. The same technology may feature in the dark detectors used by Alastor 'Mad Eye' Moody, an Auror who battled against the Dark Arts.

Take, for example, the Secrecy Sensor, a squiggly golden aerial that vibrates when it detects a fib or a lie. There is already evidence that body scanners will allow us to peer directly into the brain to find out exactly what someone believes and whether they are telling the truth. Functional MRI (magnetic resonance imaging), for example, can measure changes in the flow of oxygen-rich blood that nourishes different brain regions, and an East Coast team headed by Elizabeth Phelps, a psychologist at New York University, used this method to detect when what a person says does not always tally with how they feel.

The team studied the activity in a part of the brain called the amygdala. There is one in each half of the brain, and they become active when you are feeling fear or disgust. Even if you claim not to dislike someone or something, when you are shown a relevant picture of that person or object, the amygdalae in your brain can reveal your distaste. Perhaps the Sorting Hat can detect pure-bloods' disgust for 'mudbloods' of

Muggle descent, when it puts such pupils in Slytherin.

Another Sneakoscope that can see past our attempts to deceive measures the electrical fields generated by the activity of nerves in the brain. Lawrence Farrell, a Harvard-trained biologist, has pioneered a technique called Brain Fingerprinting, which exploits the way the brain reacts to surprising or novel information, which is measurably different from its response to material it has encountered before.

Once again, this method can see if our thoughts match our words. Electrodes connected to monitoring equipment are attached to the scalp of the person under investigation. Brain activity that reveals if something looks familiar occurs so rapidly (within milliseconds of the suspect's being exposed to material from a crime scene, such as a picture of a dead unicorn or petrified cat) that it can be detected even before a person is aware of the content of the picture. By the time he is aware, it is too late to lie about his knowledge of it.

To tell if someone is being dishonest, it may not even be necessary to peer inside his brain. There are, of course, standard polygraph techniques, where lies are detected by measuring blood pressure, breathing rate and sweating. But interpretation of such data takes time, and your suspect has to be wired up. A more elegant way to accomplish the same results may well have been found by James Levine of the Mayo Clinic in Rochester, Minnesota, working with Ioannis Pavlidis of Honeywell Laboratories. Using a high-definition thermal camera, it is possible to see that when someone lies there is a boost to blood flow around the eyes. In studies of volunteers who had committed a mock crime, measurement of concealed blushing achieved a detection rate of 80 per cent, as good as a polygraph. This phenomenon of concealed blushing may be because of the effects of adrenaline released during lying, and may have evolved to help the body prepare for an emergency. Levine speculates that blood is pumped to the eye muscles in preparation for looking around for the best escape path. Is this how the Secrecy Sensor catches red-faced liars?

The Invisibility Cloak

Another striking example of magical clothing – in the sense that it is not striking at all – is Harry's long and silvery Invisibility Cloak, left to him by his father. Supposedly woven from the hair of an ape-like creature called a Demiguise, the cloak presumably relies on the same stealthy science as *The Invisible Book of Invisibility* and the Invisibility Booster, which is activated by a silver button on the dashboard of the Weasleys' Ford Anglia. There are hints that this ability to conceal oneself from prying eyes has been around for millennia: Hades, the god of the underworld, had a 'cap of darkness', which made anyone who wore it invisible.

How this feat could possibly be achieved has been discussed endlessly. On television, invisibility is routine thanks to the use of the Chromakey (colour key) system of electronically matting or inserting an image from one camera into the picture produced by another. One camera would be trained on Harry, surrounded by a blue wall and a blue floor. Anything coloured this way can be replaced with a second image. A background picture, such as the Hogwarts Great Hall, can then be combined with the image of Harry, and the result is that he looks as if he is standing there. By wrapping himself in a blue cape, Harry could disappear completely, as the background image engulfed him.

Another theory of how the cloak might work can be found in that influential Muggle masterpiece, *The Hitchhiker's Guide to the Galaxy*, where the late Douglas Adams said that objects are made invisible by using a SEP-field. Might Harry's cloak be bathed in the self-same SEP field? Unfortunately for us, this solution is the product of a witty mind, not clever science: SEP stands for 'Someone Else's Problem'.

There are other possible explanations. A cloak of invisibility may work a spell on the brain of the beholder. Perhaps it exploits the brain's ability to paper over missing information (a skill I will discuss in Chapter 9). If a way could be found to erase information about the cloak from networks of nerve cells in the brain, the brain might fill in the missing data by extrapolating from the background around the cloak.

However, as is often the case, Mother Nature may have already solved the riddle of Harry Potter's cloak. Many creatures have perfected the art of concealment. The glass catfish, found in the Asian tropics, has transparent muscle and skin that let light pass through. In order to camouflage themselves, flatfish adjust the contrast of 'splotches' on their skin to blend into many different backgrounds in a few seconds. A peacock flounder will even provide a rough but impressive match of the squares on a chess board.

Cephalopods, such as squids, octopuses, cuttlefish and the giant squid in the Hogwarts lake, can change their appearance with a speed and diversity without parallel. An attempt to mimic cephalapod camouflage is under way at the University of Bath by Alex Parfitt. Cuttlefish have a particularly versatile way of blending with the surroundings that relies on layers of black, red and yellow colour-changing cells, called chromatophores, combined with a deeper layer of cells that reflects colours from the immediate surroundings, so they can even blend in with blue or green backgrounds. 'We are developing a gel-based system that mimics this behaviour,' explains Parfitt, 'and would like to apply it to a cover used for camouflaging large military vehicles.'

Other scientists are using electronic displays for camouflage. Canadian and German military researchers are developing a chameleon-like armoured vehicle capable of altering its appearance on the battlefield to conceal itself from the enemy. The British defence research agency QinetiQ is working on 'rugged smart skins' along similar lines. Philip Moynihan of

Caltech and Maurice Langevin of Tracer Round Associates Ltd have studied this invisibility technology for NASA's Jet Propulsion Laboratory in Pasadena, California.

Such 'adaptive camouflage' would use displays, perhaps with arrays of microscopic lasers on Harry's cloak, which would enable it to look different in response to changing scenes and lighting conditions, explains Diederik Wiersma of the European Laboratory for Non-Linear Spectroscopy in Florence, Italy. The camouflage would be able to project, on the near side of Harry's cloak, the scene from the far side of his cloak.

A typical cloak would probably include a network of electronic flat-panel display units, each containing a camera, wired into a flexible array. At every spot on the back of the cloak, it would be necessary to know the direction that the light is coming from, and its light and intensity, to set up the right image on the displays in front. The space under the cloak would also have to be illuminated in daylight conditions so Harry does not cast a shadow.

Wiersma suggests an alternative: to weave the cloak from a cross-connected network of fibre optics, the same 'light pipes' that carry telephone conversations around the planet. In this way data about a scene behind Harry would be passed along a vast number of fibres to the front of him. This would also explain why the cloak looks silvery, consisting as it does of vast number of glass filaments.

However, puzzles remain. Why does the cloak feel strange to the touch, as if water had been woven into the material? How does Mad Eye Moody manage to see through it? Perhaps the retina of his wandering electric blue eye is sensitive to a broader range of wavelengths than those handled by the cloak's fibre optic network.

NASA is also investigating the use of adaptive camouflage with an electronic 'window' that would display the great outdoors in an office that lacks a real window. Perhaps such a device could replace a conventional ceiling and provide a view of the clouds overhead, much like the ceiling of the Great Hall at Hogwarts with its enchanting view of the heavens above.

Everlasting robes

Hogwarts clothing never seems to wear out. How many references have you seen to Harry darning his robe, for example? How many times has Dumbledore replaced his purple cloak? How do the Weasleys survive on hand-me-downs? One mundane explanation is that J.K. Rowling does not want to clutter her plot lines with tedious domestic details. Another, which I find much more appealing and interesting, is that Harry wears everlasting fabric of the kind popularised half a century ago by Sir Alec Guinness in the Ealing comedy *The Man in the White Suit*.

In the classic 1951 film, Guinness played a research chemist who managed to develop a miraculous yarn that shrugged off dirt and did not wear out, much to the dismay of both management and unions of the textile mill where he worked. Now a black version of the suit is close to reality, thanks to research on filaments of pure carbon less than one ten-thousandth the width of a human hair. The 'threads' have unparalleled strength – higher than that of any other known material – and they are spectacularly damage-resistant.

These so-called nanotubes have to be assembled into long fibres to preserve their extraordinary properties, which is no easy feat. However, in recent years this problem was overcome by Brigitte Vigolo and her colleagues at the universities of Bordeaux and Montpellier, France, who created the first spinning process to make fibres from nanotubes. They assemble fibres by dispersing nanotubes – which look like soot – in a detergent solution, and then inject this solution into a flowing stream of polymer solution. In this stream the nanotubes line up into ribbon-like fibres, each of which consists of dense assemblies of trillions of aligned nanotubes. They can be strongly bent without breaking, and even tied into knots. Long and ultrastrong nanotube fibres could be used in a variety of future applications, ranging from artificial muscles to super-strong wires.

While, at present, nanotube materials are expensive for Muggles, wizards and witches who use them do not need to make so many visits to that enchanted seamstress, Madam Malkin, who makes robes for every occasion. And the persistence of these materials could explain why the used robe shop in Diagon Alley continues to thrive. But one mystery must be addressed: why does Remus Lupin, one-time Defence Against the Dark Arts teacher, wear shabby-looking robes that have been repaired in various places? The solution can be found once a month, when there is a full moon. One would, of course, expect wear and tear every time he turns into a werewolf.

Smart clothes

When it comes to creating novel clothes and materials, Muggles are only now just catching up with the accomplishments of the wizard world. There are now shirts and dresses that adapt to changes in temperature to make hot weather more tolerable, thanks to a material developed in Japan that is dotted with pores that open and close. The material consists of long chain-like molecules, built up from a number of repeated chemical units, called polymers. When the temperature rises, gaps in this particular polymer expand, allowing air in and water vapour out. In cooler weather, the gaps 'remember' their original shape and close to preserve body warmth.

In North America, scientists have bred goats that are genetically altered with spider genes so that their milk contains the ingredients of spider silk. The ingredients can be spun into synthetic silk fibres with properties approaching the real thing, paving the way for their use in artificial tendons, medical sutures, biodegradable fishing lines and soft body armour among many other possible applications. In this way, Muggles can exploit a product of 400 million years of evolution: dragline silk, found in the radiating spokes of a spiderweb, which is stronger than the synthetic fibre Kevlar, stretches better than nylon and, ounce for ounce, is five times stronger than steel.

Clothes that clean themselves with bugs that graze in their fabric are under development by other scientists. These living materials will depend on special textiles, called bioactive fabrics, made from hollow fibres containing thriving populations of harmless bacteria, or genetically modified cells, that can eat body odour and stains. Other fabrics will be self-cleaning, dust-repellent or they will continually regenerate.

Technologists are also attempting to create a convergence of computing and fashion to make truly smart clothing. One day, shoes may think, shirts change colour and skirts talk (and what tales they will tell!). Fabrics woven from fibres that contract in response to an electrical current could pave the way to a tie that can loosen itself if you get hot under the collar. Novel dyes that can change colour in response to voltage, light or heat could vary the colour of your shirt each day, decorate it with stripes or spots, or even present the latest stock market statistics.

Add the latest communications technology, and you could chat with someone thousands of miles away by mumbling to your cuffs, or download your personal details to the hat, coat or whatever item of clothing on an attractive person on the far side of a crowded room. Smart clothes could, of course, be powered by electricity captured from the mechanical power of walking. Such generators already exist in shoes. There is certainly enough room to generate a few watts of power in Dumbledore's high-heeled buckled boots, for example.

We began this chapter by considering how the new pupils at

Hogwarts are divided up into houses by the Sorting Hat. Muggle science can not only offer possible explanations of how the hat achieves this feat but can also make some predictions about what happens when pupils form rival 'tribes'. One can do this by regarding the competition between the houses at Hogwarts like a game, then attempt to capture the essence of that game with mathematics.

CHAPTER 4

The mathematics of evil

The game's afoot:
Follow your spirit; and, upon this charge
Cry 'God for Harry! England and Saint George!'

Shakespeare, *Henry V*

The rivalry between the houses of Hogwarts began not long after the school was founded more than a thousand years ago. At that time, witchcraft was still feared and its practitioners hunted down. Because of this persecution, the castle was established far from prying Muggle eyes by the greatest witches and wizards of the age. Today, the four founders are celebrated in the names of the school's houses: Gryffindor (after Godric G.), Hufflepuff (after Helga H.), Ravenclaw (after Rowena R.) and Slytherin (after Salazar S.).

Each house has its own identity, according to the Sorting Hat. The pupils who pass through Gryffindor, who include Harry Potter and a young Albus Dumbledore, are brave and daring. Hufflepuff members are loyal and hardworking, and those in Ravenclaw are wise and witty. Then, of course, there are the Slytherin pupils, who are sly, power-hungry and ambitious.

Slytherin has turned out more Dark witches and wizards than any other house. The Malfoys are members, as is Professor Severus Snape, the head of Slytherin House (portrayed in unflattering terms by Rowling, it is said, because she hated chemistry at school). Without doubt the most significant old boy of Slytherin is Harry's nemesis, He-Who-Must-Not-Be-Named, Lord Voldemort, the most evil wizard ever.

Today, mathematics can provide intriguing insights into competition, cooperation and the clash between Slytherin and the other houses. The origins of this rift are ancient. Salazar Slytherin, who believed that magical learning should be restricted to all magic families, and kept from those with Muggle backgrounds, wanted to be more selective about Hogwarts' intake. This discrimination was opposed, notably by Godric Gryffindor. Because of the dispute, Slytherin left Hogwarts (though he had enough time to set up a Chamber of Secrets that could purge the school of those he thought unworthy to study magic).

There is an informal hierarchy among witches and wizards, based on their pedigree, in the Potter books, and J.K. Rowling explores the resulting battle between a tolerant majority and a minority that is intolerant of 'mudbloods', where 'mudblood', let alone a filthy little one, is an insulting term used to describe someone who is Muggle-born.

Voldemort's father was a Muggle. However, Voldemort's veins ran with the blood of Salazar Slytherin himself, from his mother's side, and he hated Muggles with a passion, giving up his 'filthy' Muggle name of Tom Marvolo Riddle because his father had abandoned his mother. He is far from alone in holding these prejudices. The Malfoys, for example, pride themselves on being purebloods and consider anyone of Muggle descent, such as Hermione Granger, second class.

The battle for ascendancy in a society, whether between

tolerant and intolerant, or between Quidditch teams, Hogwarts houses or chess players, can now be modelled in a computer, thanks to a mathematical analysis of games developed by the computer pioneer John von Neumann and the economist Oskar Morgenstern in the 1920s and 1930s.

Game theory analyses successful strategies when the outcome is uncertain and, crucially, depends on the behaviour of the competitors. The theory is able to weigh the risks and benefits of all the strategies in a game of war, economics, or survival among self-seeking egoists. It can explain the apparent paradox of how people came to cooperate in societies, from Muggle cities to Hogwarts, when they are inclined to 'look out for number one' and nature is 'red in tooth and claw', a paradox that had troubled even the great Charles Darwin himself.

Economists are fascinated by game theory because it proposes to explain mathematically why the invisible hand of that pioneer of free trade, Adam Smith, can apparently fail to deliver the collective good. John Nash, one of the pioneers of the field, introduced the distinction between cooperative games, in which binding agreements can be made, and non-cooperative games, where no outside authority can enforce a set of predetermined rules. In the latter, he found that when all the players' expectations are fulfilled they would not want to change strategies because they would be worse off. The result is an equilibrium, now called the Nash Equilibrium, a discovery that won him a Nobel Prize.

Such work can shed light on business decisions in competitive markets, macroeconomic theory for economic policy, environmental and resource economics, foreign trade theory, the economics of information and so on. Political scientists, too, like game theory because it shows how 'rational' self-interest can make everybody worse off. In the 1970s, game theory was extended further, to biology. Now let's see what it has to say about Hogwarts.

The Prisoner's Dilemma

Interactions between individuals or groups – whether wizards, nations or companies – can be captured in a simple game called the Prisoner's Dilemma. The idea of this game is to simulate the conflicts that exist in real life between the selfish desire of each player to pursue the 'winner-takes-all' philosophy typical of Slytherins, and the necessity for cooperation and compromise in order to advance that self-same need.

It is easy to put flesh and bones on this game. Imagine that two Hogwarts pupils have been caught with a stolen wand. Because the wand is also spattered with unicorn blood, Dumbledore suspects both of having committed another, more serious, offence, of which he has no proof. There is enough evidence to punish both for possession of the wand – a deduction for the house of 50 points each – but Dumbledore is pressing to find out more about the blood, for the death of a unicorn would lose the perpetrator's house 800 points.

Dumbledore places the pupils in separate rooms and does not allow them to contact each other. He offers each of them a deal: if one informs on the other and reveals the more serious crime, his house will not lose any points for stealing the wand, and the other will lose 1,000 points. If both inform, each house will lose 800 points. If both stay quiet, each will lose 50 points for the lesser offence of possession of a stolen wand. It is reasonable for each pupil to assume Dumbledore has offered the other the same deal. What should each one do?

Imagine how, as the result of a particularly convoluted plot, the game was played by Harry Potter and one of his enemies – and Malfoy's awful chum – Vincent Crabbe. From the point of view of Crabbe, there are two things that Harry can do: cooperate (keep quiet) or defect (confess and implicate the other). Crabbe can then reason as follows: if Harry cooperates, and then if I cooperate, both of our houses lose 50 points. But if I defect, Slytherin loses none while Gryffindor loses 1,000.

Thus I should defect. But what happens if Harry defects? Then if Crabbe cooperates Slytherin will lose 1,000 points, but if he defects then each will lose 800, so Crabbe should defect, too. Thus the 'rational' thing for Crabbe to do is defect. Since the payoffs are the same for Harry, he should come to the same conclusion. Both will be led by logical 'reason' to defect, and so they will end up losing 800 points each, and be severely punished, even though they would have lost only 50 if they found a way to cooperate with each other and maintain their silence.

The Prisoner's Dilemma exercises mathematicians, social scientists and biologists because it helps to elucidate the details of a widespread problem: how individual ambition can lead to collective misery. In our example, the outcome is influenced by many factors: if both pupils had belonged to the same house, they may have been more trusting; if they were never going to meet again, there would be no reason for them to cooperate, and so on. But in real-world situations it is often more likely that the parties will encounter one another in the future. Then different strategies emerge in this repeated, or iterated, Prisoner's Dilemma.

One influential figure in the field is Robert Axelrod at the University of Michigan, who held a worldwide tournament in 1980 for computer programs to play the Prisoner's Dilemma in an attempt to uncover the best. The winner was Tit-for-tat. As its name suggests, it involves cooperating in the first round, and then doing whatever an opponent does in successive rounds. While it is a 'nice' strategy, in that it signals willingness to cooperate, it is ultimately not too clever. Highly complex strategies can seem incomprehensible to an adversary and remove any incentive to cooperate. Tit-for-tat's great success is its simplicity, and its discovery as a successful strategy sends an optimistic message to those who fear that human nature is founded on greed and self-interest alone, typified by the behaviour of the Malfoys. When it works in human society at large, the Tit-for-tat strategy means that a successful entrepreneur may be an opportunistic seeker of cooperation rather than a ruthless operator. Cooperation can emerge in a society according to the principle 'give and you shall be given'. Nice people, such as the members of Gryffindor, do not have to finish last.

However, in the real world, nice guys can and do finish last. The emergence of cooperation can be triggered when any one of the following conditions are met: the 'players' repeatedly encounter each other; they recognise each other; they remember the outcome of previous encounters. But there are other factors that must be taken into account, from the chance of an encounter between players to mistakes, when attempts to cooperate are perceived as defection, to the probability that genetic factors that shape behaviour are passed from generation to generation. With such real-world uncertainties in mind, Robert (now Lord) May of Oxford University pointed out in 1987 that Axelrod's groundbreaking work is highly idealised and unlikely to apply wholesale to the real world.

Attempts to find out how Tit-for-tat copes with these complications were made by Martin Nowak, now at the Institute for Advanced Study, Princeton, and Karl Sigmund, working at the University of Vienna. They found that by adding errors, which might be caused simply by the all-too-human tendency to make mistakes, then Tit-for-tat is no longer the supreme strategy because it is unforgiving: once two Tit-for-tat players start defecting, they will continue to do so. By including the element of uncertainty, individuals can develop new strategies. The addition of a little randomness to behaviour allows 'forgiveness', and a chance to test the behaviour of another player. One strategy that takes this into account is 'generous tit-for-tat', where the randomness can break cycles of mutual backbiting. Another, more successful, is called Pavlov, which can be summed up by the maxim: 'If it ain't broke, don't fix it (and if you lose, change strategy).'

Uncertainty does, however, allow cooperation, and the optimistic message of Axelrod's work remains.

Another elaboration of the Prisoner's Dilemma was carried out in research by Martin Nowak and Robert May, who modelled the game in two dimensions, investigating what happened to a chessboard of players. By introducing geography into the Prisoner's Dilemma, cooperators and defectors can exist side by side, not as individuals but as groups. In the real world, this means that diverse populations of hosts and parasites or prey and predators can persist in communities, despite the instability of their interactions.

This explains why the enclosed society of Hogwarts is made more stable by giving a dormitory to each house, rather than making the entire student body sleep in one giant, randomly mixed hall. Mathematical models suggest that when cooperators, such as Gryffindor, meet other cooperators, they tend to prosper. The non-cooperators, such as Slytherin, don't completely take over, even though in any given exchange with a cooperator they do well, because defectors do poorly when they interact with each other, while cooperators thrive and can even 'convert' some defectors to the better payoff from reciprocal cooperation.

But this is still a very idealised model. Large numbers of people cooperate in societies where, unlike the neighbours on a chess board of players, they often encounter individuals they will never see again. Yet they give to charity, take part in political movements and conform to social norms. Given the number of one-shot interactions between real people, why doesn't society just fall apart? Experiments with human guinea pigs suggest that punishment is the answer.

Ernst Fehr of the University of Zurich and Simon Gächter of the University of St Gallen, both in Switzerland, devised a financial game in which participants – all of whom were strangers – had to decide whether to commit their individual resources to a common pot or hold back and reap the benefits of the community spirit of the others. This game is an attempt to solve a long-standing social dilemma, the 'tragedy of the commons' or the 'public goods game'. In this game people are asked whether they would contribute money to a public pool that is doubled by the experimenter and then distributed evenly, irrespective of whether or how much they contributed. The group as a whole will have the greatest profit if all its members contribute maximally, but the first player who defects and does not contribute will profit most personally. This public goods game is regarded as the paradigm for many human problems – for instance, our inability to look after the environment and sustain the global climate. In the world of Hogwarts, one example of the public goods game could be a communal whip-round to buy all the racing brooms for a house.

In the Swiss experiment, the 240 players were shuffled so that no participant encountered anyone else more than once and so that no one who took part could develop a reputation for being generous or mean. Without a financial punishment being applied to those who did not make public-spirited investments and only exploited the generous nature of others, cooperation rapidly foundered over six rounds as other players punished the free riders by withdrawing their own cooperation. Society fell apart. 'Our experiments show that if group members can only punish by withdrawing their cooperation, cooperation by all members quickly unravels,' says Fehr.

When an explicit punishment was enforced against free riders, however, the common good prospered; in this case, more than 90 per cent of the players contributed more money. 'If explicit punishment of non-cooperators is possible, cooperation emerges even in one-shot interactions when there is no repetition,' says Fehr. 'It is important that punishment can be targeted at the free rider and is explicit, that is, not merely the withdrawal of cooperation.'

The most successful strategy used 'altruistic punishment', in which there is a financial cost to individuals who mete out the punishment, making them slightly worse off overall. The researchers found that, despite the cost to them, the players derived psychological, rather than material, benefits from this strategy because they were so annoyed and resentful when exploited by others. Thus, more than simple self-interest motivates and sustains cooperation; altruistic punishment is one component of the 'glue' that holds society together.

The Hogwarts caretaker, Argus Filch, talked longingly about being allowed to once again suspend guilty students by their wrists from the ceiling for a couple of days. Indeed, he kept a collection of chains and manacles for that purpose, just in case. His emphasis on inflicting pain does not sound very altruistic, since there seems to be no cost to him, and indeed he may even find it pleasurable. However, the Hogwarts that Harry is familiar with often deducts points from a pupil's house for bad behaviour, just as surely as it rewards triumphs. This does lead to altruistic punishment, since members of the House resent and criticise colleagues whose bad behaviour loses them points, causing an overall drop in morale. Hermione Granger told off Ron and Harry for wandering around at night by reminding them that they put Gryffindor points at risk. After more nocturnal wanderings lost the house 150 points, Harry went from being one of the most popular pupils to being one of the most hated. 'Pupils punish each other by ridiculing their peers for violations of group norms. Most peer punishment of group norm violations is altruistic,' says Fehr.

Another example of altruistic punishment by a group might be the case where a Quidditch team doesn't allow a star to play, because the star broke some team rules. Detentions, another punishment at Hogwarts, also look rather altruistic: a teacher has to waste her own time supervising a detention; and when Harry had to help search the Forbidden Forest with Hagrid in the dead of night, Hagrid was also put at risk. 'Often superiors don't like to punish. In fact, punishing others is often not convenient. Our emotions help us overcome this inertia in punishing others – and then it counts as altruistic punishment,' says Fehr. 'So even the punishment executed by superiors – although formally they have the right to punish (like policemen or teachers or superiors in an organisation) – may count as altruistic punishment.' Unwittingly, J.K. Rowling has illustrated how the Utopian society, in which it is in everyone's interest to cooperate without the threat of sanction, is just that, an impossible dream.

Good Samaritans

The most obvious explanation for the evolution of cooperation is relatedness: cooperation emerges if there is a sufficiently good chance that you will encounter a brother, cousin or someone who is genetically related to you (the exception, of course, being Harry and the Dursleys). If you happen to have a trait that makes you cooperate, chances are that the same trait can be found among family members. This explains why in colonies of insects, there are infertile workers, soldiers who sacrifice themselves for the greater good of the colony and so on. Humans are peculiar because, unlike animals, they have solved the problems of organising large-scale cooperation among non-relatives, such as between the players in a Quidditch team. But we also help people we don't know and whom we may never see again. What motivates a Good Samaritan to help a stranger he is unlikely to encounter again, as is uniquely the case in human society? For example, when Harry undertook the second task in the Triwizards Tournament he not only rescued his friend Ron from the murky depths of Hogwarts lake and the merpeople, but also a girl whom he had never met before, Fleur Delacour's sister Gabrielle.

One explanation is that the prospect of encountering the same people in the future allows cooperation to be built on the basis of reciprocity, as in a strategy of Tit-for-tat. Another idea, put forward by Martin Nowak and Karl Sigmund, is that if you advertise you are a good guy, you boost your chances of being helped by a Good Samaritan at some future date. This idea, which was successfully tested by Manfred Milinski of the Max Planck Institute of Limnology, Ploen, Germany, with Claus Wedekind, is summarised in the lines from Tom Lehrer's parody of the Boy Scouts, entitled 'Be Prepared': 'Be careful not to do/Your good deeds when there's no one watching you.' The gossipy society of Hogwarts (where Harry could not keep secret the fact that he was training as a Seeker for Quidditch, or details of his battle with Voldemort in the dungeons, for example), must have given this strategy for good behaviour a boost. Reputation is another factor that holds societies together, according to work by Milinski on the public goods problem that complements that of Fehr.

Another explanation of Good Samaritans has come from game theory computer simulations by Rick Riolo, of the University of Michigan Center for the Study of Complex Systems in Ann Arbor, and his colleagues. Rather than focusing only on individuals having similar genes, their computer study shows that Good Samaritans are also interested in individuals with some recognisable common trait, whether Hogwarts robes, school ties, club memberships, tribal customs, religious creeds or memes (units of cultural transmission, from fashions to legends). These are all tags that induce cooperation, like the house symbols, from Gryffindor's lion to Slytherin's serpent. 'We find that if individuals are generous to others who are very similar to themselves, this, too, can sustain cooperation. And this works even if the basis for judging similarity is a completely arbitrary characteristic,' says Riolo. Perhaps it is no coincidence that to find out how much one has in common is among the first delights of falling in love.

Significantly, the study shows that tolerance cycles over time so that, when a few intolerant individuals – such as the

Slytherins – emerge in a tolerant society with a different cultural or social makeup, the newcomers receive much more help than they give. Because the original Slytherins in Hogwarts would have received more benefits without paying the cost of helping, they would increase their numbers, and a wave of intolerance consequently swept through the population, along with a drop in overall cooperation.

However, after rapid growth in numbers the Slytherins themselves would become the dominant group, which is exactly what happened when Voldemort and his followers, the Death Eaters, rose up about a decade before Harry was born, paving the way to the dark days when Muggles were killed for fun. This in turn would lead to an increase in overall cooperation since the Death Eaters, Slytherins and others like them, despite being intolerant, do help one other, until this new dominant culture is in turn challenged. The rest is described in *The Rise and Fall of the Dark Arts*. Though it is foolish to argue that the subtleties of real life can be boiled down to the vagaries of a virtual population in a computer, such computer simulations are striking.

Gobbledegook and game theory

Game theory can also shed light on the many languages encountered by Harry Potter. For example, there is Mermish, Gobbledegook and Troll. Then there is Parseltongue, used to chat with snakes. While there are no fossils to tell us how language came into being, the reason that humans communicate with strings of words rather than grunts can now be explained with evolutionary game theory, a field that combines game theory with the insights of Charles Darwin, pioneered by John Maynard Smith of Sussex University and the late Bill Hamilton of Oxford University.

Language is a game for more than one player, according to

Martin Nowak. Some time between seven million years ago, when we shared our last common ancestor with chimps, and 150,000 years ago, when anatomically modern humans emerged, the game of language must have come into being.

In his first study, conducted with David Krakauer, Nowak began by examining the 'primordial soup' of language present throughout the animal kingdom, whether in the form of primitive signalling between cells, the dance of bees, territorial calls and birdsong. They showed how natural selection guides the three steps in the evolution of human language, from sounds to words to a 'protogrammar' spoken by our distant ancestors.

The first step in language evolution was to link noises with the world about them. Early humans had a few specific utterances, from howls to grunts, that became associated with specific objects. This category of primitive communicators includes Trolls, which are so dim that they are classified as beasts in the wizarding world. Fred Weasley observed how Trolls set up associations with a point and a grunt.

Crucially, these associations form when information transfer is beneficial for both speaker and listener, underlining how the evolution of cooperation, discussed earlier, is crucial for language to evolve. However, this first stage of language quickly runs into problems in any society that has the capacity for advancements, for more signals will soon be needed as it attempts to describe more objects. There is a limit to the number of simple verbal signals because, as the number increases, so too does the probability of misinterpreting and muddling them. It is difficult to distinguish one grunt from another, particularly if there are many in use. In essence, the more concepts that have to be communicated, the more sounds that are needed to do so, the closer the sounds will be to each other, and the greater the risk that this repertoire will become confused.

In the second step of language evolution, humans leave

animals behind and overcome this 'error limit' not by forming more sounds but by combining a small set of easily distinguishable sounds into words. Meaningless vowels and consonants such as e, l and f can be assembled into a meaningful word: 'elf'. Nowak's team showed mathematically how such word formation enables a language to convey an essentially unlimited number of objects, and explained why, although there are many more complex languages, they all are composed of a few sounds.

Trolls are pea-brained and, like animals, use non-syntactic communication, so that a single grunt – a 'word' – might be used to express a scenario, such as 'There's a centaur galloping through the Forbidden Forest.' The syntactic communication we rely on uses words for the individuals, actions and relationships that contribute to the event, giving it greater expressive power.

The last step in language development is the incorporation of grammar, which enables words to be combined in essentially an unlimited way. Nowak showed how simple grammar rules evolved to reduce mistakes in communication, shuffling the 'atomic units' of language to create an infinite number of meanings. Although this form of communication is in some ways more abstract and less immediate than howls, snorts and so on, it is much more flexible. For example, in a proto-language the phrase 'murtlap [a ratlike creature] bites man' would be described by a different sound from 'man bites murtlap' and would have to be learned separately. In contrast, in a grammatically organised system, a rare but important event can be easily described by combining words in ways that need not be learned beforehand.

But syntax comes with a cost, for grammar requires a degree of mental exertion. That poses the question: under what conditions are communicators encouraged to shift from non-syntactic communication of animals to the syntactic communication used by humans? Nowak and colleague Joshua

Plotkin, working with Vincent Jansen of Royal Holloway University of London, found that in an environment of sufficient complexity, where survival depends on important information being quickly dispersed, its benefits exceed its costs and syntactic communication should win: language should come into being.

Thus one might argue that language and other complex mental capabilities might be driven by the need to cooperate in the face of potential defectors, free riders and so on. This could then lead to an 'arms race' in cognition, as defectors then try to get smarter so they can lie or otherwise defeat the mechanisms cooperators use to identify one another, then cooperators discover means to defeat those mechanisms and so on. Perhaps eternal struggles like that between Slytherin and Gryffindor have helped drive the evolution of our brains.

Another implication, says Jansen, is that having syntax pays off only if there is much to discuss. In a simple environment, presumably one like that enjoyed by mountain Trolls, where all there is to gossip about is where the next meal is coming from, how to wield a club and so on, there is no need for grammar; a few grunts will do. (Muggle myth certainly suggests as much: the Norse settlers of Shetland used the expression *trollmolet* to mean troll-mouthed, or surly.) The structure of human society could, Jansen argues, be the driving force behind the emergence of grammar. He contrasts the language required to make a potion with that required to parlay the gossip at Hogwarts – for instance, speculation about who would accompany Hermione to the Yule Ball.

'In terms of linguistic complexity, recipes are very simple. Try using a foreign language recipe book and, so long as you know what the words mean, it is remarkably easy. If you can work out the words *pulverised spider legs*, *cauldron* and *stir* it probably means put the pulverised spider legs in the cauldron and stir,' says Jansen. But if you overhear a conversation and hear the words *ball*, *Hermione* and *stranger*, it could either mean that Hermione is asking a stranger to a ball or is asked by a stranger to come to a ball. Here word order is important. Indeed, the meaning of the word 'ball' is also context-dependent. This sort of conversation uses the grammatical structure of language to the full.

Game theory is still struggling to accommodate the Jarvey, a ferret-like creature that confines itself to short, mostly rude, phrases but lacks real conversation. Nonetheless, it seems that our language was probably born of soap opera: the structure of human society is so complicated that we had to evolve the gift of the gab. This work suggests that syntax should not be regarded, as it often is, as a hallmark of human intelligence. Other creatures, such as the Acromantula spider, elves and merpeople, sirens, selkies and Merrows, also evolved language. They, too, must have had a lot to say for themselves.

CHAPTER 5

Owls, snails and Skrewts

It was the owl that shriek'd, the fatal bellman,
Which gives the stern'st good night.

Shakespeare, *Macbeth*

The owls fly through a starless night, black ink on black paper, shadows of shadows. They glide on nameless winds, seeking currents and eddies. Circling, they form a column of silently beating wings and staring eyes. The column sets up tremors in the breeze. The tremors bring more owls, and they join the circling around the West Tower of Hogwarts. Something is in the air tonight.

Whenever owls gather in large numbers in the Harry Potter books it signals front-page news, momentous events and headline happenings. After Voldemort's power was broken in his first encounter with Harry, for example, Muggle bird-watchers had a field day observing these usually nocturnal creatures fly hither and thither.

They are to Hogwarts what the post is to the Dursleys and other Muggles. Any self-respecting witch or wizard needs an owl. Parcels, mail and other messages can be tied to a leg, or carried aloft by claw or beak. The variety of swooshing feathered messengers is vast. Harry's Hedwig is a snowy or ghost owl, for example, while the Weasleys' moribund Errol is probably a great grey, also known as a spectral owl. Even colourful tropical birds have been known to make the odd delivery.

These birds are among a wide range of extraordinary real-world creatures and their appendages that make an appearance in the Harry Potter series. Think of that giant squid lurking in the Hogwarts lake, or those orange snails in the Magical Menagerie – were they something to be welcomed or feared by the pupils of Hogwarts? As for the blast at the end of Hagrid's Skrewts, surely there is nothing in the real world like it?

Thanks to efforts by Muggle zoologists, ornithologists, entomologists and marine scientists, we know a lot about these fauna and their extraordinary capabilities. The giant squid is one of the most elusive and fascinating of them all, which scientists are still struggling to understand. The Skrewt uses natural technology that the Nazis came to reinvent to bomb Britain. And those orange snails have an intriguing pharmacological tale to tell, one that can reveal both good and evil reasons to keep them at Hogwarts.

The owl post

Perhaps the most pressing question of all is whether owls really are up to the job of delivering mail. First and foremost, can they get very far? And how much can they carry? The answers probably depend on which kind of owl you are talking about. These birds are found across the planet and number more than 300 species. Some of this variety is reflected in Eeylops Owl Emporium.

The smallest owl is the sparrow-size elf owl from Mexico and the southwestern part of the United States. This is probably the species that Pigwidgeon belonged to, being no bigger than a golden snitch. Even its name suggests its size and echoes Pigwidgin, a term used for a fairy or dwarf. Tiny twittering Pig (widgeon) struggled even to carry a single letter and was small enough to fit into the palm of Harry's hand. At the other end of the size spectrum is the Eurasian eagle owl, owned by Malfoy, which has a wing span of nearly two metres and is strong enough to carry heavy parcels.

Despite their routine use as messengers by witches and wizards, owls do not make good pets, says Colin Shawyer, director of the Hawk and Owl Trust in Britain. Like any wild creature, the birds fare best in the wild. It is conceivable that Ron Weasley could look after Pig who, like elf and pygmy owls, probably preys on a variety of insects, spiders and other bugs. However, other species would be hard to care for since they often dine on animals that are almost as large as themselves. Malfoy would find his feathered messenger particularly demanding to feed without the help of magic, for eagle owls have been seen attacking golden eagles, foxes, herons and dogs and there is apparently a report of a large Siberian bird taking an adolescent wolf. Fortunately, however, most owls at Hogwarts seem smaller than this. We can obtain clues to their diet because the birds often swallow their prey whole – fur, feathers, bones and all – and then, after a few hours, regurgitate the indigestible parts as a pellet. The straw of the Owlery is strewn with the remains of mice and voles, suggesting that Hogwarts relies on medium-sized Strigiformes.

Wise owls and wizards

The appearance of owls in the enchanted world of Harry Potter reflects a fascination that dates back many thousands of years. Rock paintings of the birds have been found in disparate locations, from Australia to Vallon-Pont-d'Arc in southeastern

France, where an engraved owl perches among some paintings that date back 30,000 years – no doubt the product of a fur-clad Michelangelo.

The link between owls and folklore is equally ancient. The bird was the symbol of ancient Athens, for example, and the phrase 'to send owls to Athens' is the equivalent of sending coals to Newcastle in Britain today. The silver four-drachma coin bore the image of the owl as a symbol of the city's patron, Athena, the Greek goddess of wisdom. Hence the idea of owls as messengers of wisdom – in the form of letters at Hogwarts. Another link with Hogwarts is in T. H. White's classic *Once and Future King*, where Archimedes the owl perches on the wizard Merlin's shoulders. Perhaps it is no wonder that in that other children's classic, *Winnie the Pooh*, problems usually prompt Pooh to seek Owl for advice since, 'If anyone knows anything about anything, it's Owl.'

In some cultures, owls serve to represent human spirits after death; in others, they are spirit helpers that enable humans (often a shaman, or medicine man) to harness their supernatural powers. Among some native groups in the Pacific Northwest of America, the nocturnal birds served to bring shamans in contact with the dead, see at night, or locate lost objects. Some Native Americans wore owl feathers as magic talismans, while, in Japan, owl images and figurines were placed in homes to ward off famine or disease.

Although everyone likes to get mail, not everyone would have welcomed the soft *flump* of an owl on their bed in days gone by, when they were seen as bearers of bad tidings by some. The Roman statesman Pliny the Elder wrote that owls foretell only evil and are to be dreaded more than all other birds. There is a suggestion that the word 'owl' is linked to the Icelandic *uggligr* which means 'fearful or dreadful'.

By the Middle Ages in Europe, the owl's eerie call filled people with foreboding, and it became associated with witchcraft. To be 'owl-blasted' was to be bewitched. Shakespeare wrote of the owl as the 'fatal bellman'. Throughout India, owls are construed as bad omens, messengers of ill luck, or servants of the dead. Until recently, owls were nailed to barn doors in Britain to ward off lightning and the evil eye.

It is the nocturnal habits of most owl species that probably led to their being attributed with occult powers. This was underlined by their virtually noiseless flight, due to the velvety surface and, in some owls, the unique serrations on the leading edges of their flight feathers. The bird's staring eyes and superb night vision may be responsible for its connection with prophecy and its reputation for being all-seeing.

Most people associate owls with a hoot, which is typical of the sound made by the tawny. However, their calls range well beyond the humdrum hoot. Even the tawny has a greater vocal repertoire, delivering shrieks as well. The European scops owl is said to have bell-like chimes; the barn owl a strangulated screech; eagle owls a deep, resonant hoot; and pygmy owls melodic trills. India's forest eagle owl can make a moaning hoot and a blood-curdling shriek, reminiscent of a woman stricken with grief. Locally, it is known as the devil bird.

The world of owls

Not all real-world owls would make good mail birds because most species are sedentary, according to Colin Shawyer. Candidates to haul letters, packets and parcels any distance for Hogwarts must come from the few species that migrate. Fortunately, Britain is blessed with a high proportion of migratory species. Two out of the five native UK species – the long-eared and short-eared owls – fly distances of up to hundreds of miles to and from the Continent, enough to reach the Durmstrang Institute or the Beauxbatons Academy of Magic.

Most owls could deliver mail at night, which is handy so

they don't draw the attention of Muggles. They owe their awesome night vision to several adaptations; owl eyes are large and tubular, rather than round, giving them a relatively large cornea in proportion to the overall size of the eye, allowing more light to enter. The lens of an owl's eye is large and convex. The light passes through the pupil, which can be opened so wide that virtually no iris is visible. An owl's retina is packed full of light-detection cells, called rods, which are far more sensitive than other light-sensitive cells, called cones, at low-light levels. The phenomenal light-gathering properties of an owl's eye is further enhanced in many nocturnal species by a reflective layer behind the retina, the tapetum lucidum, which gives the retina a second chance to detect light by reflecting back any that made it through the retina initially.

The eyes of tawny owls, which are the best-developed of all the owls, are about 100 times more sensitive at low light levels than our own. To put this into perspective, the World Owl Trust points out that a tawny owl can locate prey just a few yards away by the light of a single candle about 1,700 feet away.

But even if these birds could see the distant lights of Hogwarts on a bleak winter's night, could they possibly remember where to take a parcel and figure out how to get home? There has not been a systematic study of owl brain power, though the memory of barn owls has been probed by Eric Knudsen at Stanford University. The working memory (the kind we use to store phone numbers before dialling them) of the birds is excellent and is located in a part of the forebrain called the archistriatum. Old owls have a reputation of being wise and, according to Knudsen's work, they would be good at remembering a new destination for Hogwarts' mail.

Knudsen has tested their ability to remember the location of savoury prey, such as a cricket, while thinking about something else. Owls determine the position of prey by measuring exactly when the sound from their intended meal reaches their ears. A sound that originates at the extreme left of the owl will arrive at the left ear about 200 microseconds (millionths of a second) before it reaches the right ear.

Because owls' eyes are essentially fixed (they only make small eye movements) in their sockets and cannot rotate, the birds turn quickly in the direction of a sound. They have deceptively long, flexible necks that enable them to turn their head through a range of 270 degrees.

Barn owls, which have a much more sensitive auditory system than humans, and possibly any animal, quickly locate a chirping cricket or a squeaking mouse by integrating two neurological 'maps' in their brain – one linking location to sound, the other to vision.

To probe the role that owl memory plays in linking the sight and sound of a prospective meal, Knudsen fitted the owls with prism spectacles that shifted their visual field. As a result, their response to sound also shifted to compensate so that the location of the sound matched what they saw. A young owl fitted with the spectacles thus learned that the sound of a cricket coming from the right demands a straight-ahead glare, whereas a sound from straight ahead necessitates a head movement to the left. The bird's skewed mental map of sound space, in the midbrain, is shaped by visual experience and merged with a visual map of the surroundings in another part of the brain called the optic tectum.

But the ability to cope with the prism spectacles depended on the age of the birds: young owls learned new mental maps of sound space far better than adult owls did. Adult owls reared with prism spectacles that had subsequently been removed could adapt to the spectacles successfully when they were introduced again. As adults, they quickly remembered how to accommodate the prism's zany view. But adult owls introduced to the glasses for the first time never learned to adapt, proving you can't teach an old bird new tricks. This experiment suggests that young birds have sufficient flexibility in brain architecture to lay down new networks of nerve cells needed to tackle novel tasks.

To learn a new location the bird must also forget the old responses that are no longer appropriate. To find the seat of forgetting, Knudsen focused on the ICX region in the midbrain of the barn owl, one rich in a certain class of nerve cells that damp down, or inhibit, activity, called GABAergic nerve cells. He reasoned that, by blocking these cells, the forgotten map of behaviour would move to the forefront, replacing the active map. And, indeed, when the GABAergic cells were blocked, cells in the owl's brain that had been showing the learned map suddenly began to show the forgotten map. Knudsen concluded that the normal map had not been lost, but that it and all the information it contained had remained intact, and had been temporarily suppressed, or forgotten.

Are owls smart?

So far we have established that young owls can learn new tricks, that they can suppress memories that are likely to be confusing, and that they can learn associations and retain them. While that ability suggests that they could learn to deliver something to a given destination, it seems to fall short of establishing that they are particularly wise. For example, could owls communicate with witches and wizards? Could they demonstrate planning and conscious thought? Or could the birds understand that some events have taken place in the past and use this knowledge to guide how an individual should behave in the future?

There is much evidence of intelligence in the bird world. Grey parrots have been able to master tasks that rival those achieved by chimpanzees and apes, according to Irene Pepperberg, who works at MIT and Brandeis University. One of her youngest charges, Griffin, began by combining objects in specific orders, and doing the same with vocal labels. Such behaviour was once thought exclusive to humans, great apes and monkeys, according to Pepperberg. 'The fact that we are finding this in animals so far removed from primates is exciting,' she says.

Griffin can order bottle caps and lids in certain patterns and speak word combinations that roughly correspond to human language (such as 'green birdie', or 'do you want grape?'). Griffin is at an early stage in label acquisition; he can name about a dozen different objects or materials and is learning his colours and shapes. Pepperberg's oldest and most accomplished bird, Alex, can label more than fifty exemplars, including seven colours, five shapes, quantities to six, and three categories (colour, shape, material), and use 'no', 'come here', 'wanna go X' and 'want Y' (X and Y are appropriate location or item labels). Alex combines labels to identify, request, comment upon or refuse more than 100 items and to alter his environment. He processes queries to judge category, relative size, quantity, presence or absence of similarity or difference in attributes, and shows label comprehension. He can be presented with a collection of seven items and respond correctly to a sentence as complex as 'What shape is green and wood?' Alex 'thus exhibits capacities once presumed limited to humans or nonhuman primates', Pepperberg says. Rather than simply associating the patterns of labels with food, she believes Alex really understands what he is saying.

Other experimental evidence suggests that birds are not bird-brained. In mythology the crow family, or corvids, which includes jays, magpies, ravens and jackdaws, is often billed as being smarter than other animals, even humans. In the mid 1990s Gavin Hunt witnessed New Caledonian crows in New Zealand using hooked twigs as tools to capture food. Then it was discovered that western scrub jays seemed to have a type of memory that was previously thought unique to humans.

By giving birds perishable and non-perishable food to store, and then allowing them to recover the items either a short or long time later (when perishable food was still fresh, or rotten, respectively), Nicky Clayton and Tony Dickinson of Cambridge University demonstrated that the jays were able to remember what a past experience was, and where and when it occurred. The birds could recall not only where they had stored food – by covering with a leaf, stuffing it into a crevice or burying it – but the type of food and the relative time that had elapsed since storage. Following up this work, Nicky Clayton and her husband, Nathan Emery, found the first evidence that the same jays are able to carry out mental time-travel – that is, use their memories of past experiences and to plan for the future.

The intellectual prowess of the jays had begun to niggle Clayton during her lunch hours while working at the University of California, Davis, when she noticed there was fierce competition between the birds for crumbs and other left-overs from meals of students and staff. To protect their hard-won scraps, the birds would hide their winnings. But some scrub jays went even further – returning to re-bury the treasure when their rivals had left the scene. Presumably, the birds did this to prevent pilfering, she reasoned.

With Nathan Emery, she tested this in a series of trials to see if the birds hide food according to their previous experiences as a thief. Clayton and Emery discovered that if the scrub jays are observed storing food by another jay, birds with a criminal past subsequently move their store to a new location when they have the opportunity to return to their stashes in private. Those more innocent of the ways of the world are far more trusting: they leave their food in their original hiding place, even if they know that another bird has seen them put it there. In other words, it takes a thief to know one: thieving jays watch their backs more than honest birds because they know, from their own experience, that their food could be stolen.

'The experienced scrub jays appear to have projected their own experience of pilfering to the future intentions of another bird. This suggests that the jays possess some of the hallmarks of theory of mind,' says Emery. This is a remarkable discovery, because the ability to read another individual's intentions, beliefs and desires develops around the third year of life in humans and had yet to be demonstrated convincingly in animals. 'To our knowledge this is the first experimental demonstration that a non-human animal shows elements of mental time travel,' says Clayton. 'The experienced jay must be using some cognitive ability to perform this behaviour, not necessarily in the same way as humans do. Nonetheless, this is something only chimpanzees and other great apes have been suggested to do.' If true, scrub jays will be the only non-human species that has been shown so far to possess a theory of mind. The research provides an imperfect but revealing mirror with which to judge human skills and abilities.

The discovery that jays are avian Einsteins raises a question: are owls as smart as jays? 'From casual observation . . . I am sure that they are not as smart as a jay, crow or parrot,' says Knudsen. The animal trainer for the first Harry Potter film was less polite about some of his feathered charges: 'Snowy owls are real dim. They've got their niche in the wild so well organised, they don't need much imagination.' Like so many areas of science, it seems that more research is necessary to see if owls are up to the job of mail delivery. However, as with many aspects of the magical world of Harry Potter, the feat is certainly not beyond the realms of possibility, given the talents of other birds. And, as will be illustrated by other examples in the next chapter, it may even be possible to use genetic modification to turbocharge an owl brain.

Giant squid

Before my eyes was a horrible monster, worthy to figure in the legends of the marvellous. It was an immense cuttle-fish, being eight yards long. It swam cross-ways in the direction of the Nautilus with great speed, watching us with its enormous staring green eyes.

Jules Verne, *Twenty Thousand Leagues under the Sea*

Another creature that looms large in the Harry Potter series is the giant squid that lurks in Hogwarts lake and enjoys basking in its warmer shallow waters, where it is sometimes fed toast by the pupils. Like so much else in the books, it has its origins in folklore. There are many tales of bottomless pools inhabited by monsters: the 'Black Mere of Morridge' – near Leek, Staffordshire – where a mermaid lured travellers to their deaths; at Ellesmere, Shropshire, Jenny Greenteeth lurked under the weeds of a stagnant pool to catch unwary children; and Lake Avernus near Pozzuoli was once thought to be an entrance to the Underworld. No doubt these tales served a useful purpose – to keep children away from water and reduce the number of drownings.

There is also a real-world version of the Hogwarts squid that is just as mysterious. Norwegian legends tell of krakens, marine creatures so large that, from a distance, they resemble islands. The bishop of Bergen in 1753 described an immense sea monster 'full of arms' that was big enough to crush the largest man-of-war. Remarkably, however, this giant squid is real. Called *Architeuthis* – the Greek for 'chief squid' – it is one of the great mysteries of life on earth.

Lurking fathoms below the surface of the ocean, the creatures are so elusive that the adults have never been seen alive. Occasionally these leviathans, which are the world's largest and most advanced invertebrates (animals without backbones), become tangled in fishing gear in the inky depths and get hauled to shore. So far, the only live examples of the leviathans ever studied have been just half an inch long, captured for the first time by Steve O'Shea of the National Institute of Water and Atmospheric Research in Wellington, New Zealand. He found these larvae at a depth of just 20 feet around 150 miles east of New Zealand, and DNA tests at two laboratories confirmed them to be *Architeuthis*.

O'Shea is now trying to find a way to grow them in the laboratory. The larvae have some growing to do. Estimates of lifetimes have had to come from studies of growth rings in statoliths, calcified structures found in the head of the monster, which suggest spectacular growth rates: from smaller than a shrimp to their full adult length within around eighteen months. The monster typically grows to at least sixty feet long, weighs one ton, and has lidless eyes the size of dinner plates (at ten inches, they are the largest in the animal kingdom). Some of its nerve fibres are so big they were initially mistaken for blood vessels. The size of the creature has long been inspirational, most famously to Jules Verne, who described how the giants attacked the submarine *Nautilus*.

Architeuthis possess the same chromatophores that provide smaller squids with their amazing ability to change colour. Like its close relative, the octopus, the squid has eight arms. But unlike the octopus, it has two longer tentacles with suckers on the ends that are used to grab prey. All ten appendages of the squid are, as Jules Verne notes, 'fixed to its head', and are arranged in a circle around its mouth – 'a horned beak like a parrot's' that is as big as an outstretched human hand.

Paradoxically, little is known about these huge creatures even though they appear to be common: captured sperm whales usually have squid beaks in their stomachs, the beak being the only indigestible part of the leviathan. Clues to where these encounters take place have come from New Zealand. Local fishermen and scientists have developed commercial fisheries, going after such exotic fish as hoki, ling and the orange roughy, around Chatham Rise, a rocky plateau the size of Texas that lies half a mile or so beneath the waves. There, they would haul up dead giant squids that had been feeding on dense schools of fish. The adults are thought to live at depths of between 900 and 3,000 feet, with most found at around 1,500 feet. Thus the Hogwarts lake must be exceptionally deep and it must be linked to the sea in some way, since these are marine creatures.

The fact that Hogwarts *has* a resident squid may provide a few clues to the school's whereabouts. Martin Collins of Aberdeen University points out that a number of specimens of *Architeuthis* have been caught or stranded around the coast of Scotland, where Hogwarts is assumed to be: from North Uist on the west coast to Shetland and Bell Rock, off Arbroath, on the east.

Could a marine environment be found in a Scottish lake? Some Scottish lakes may have been saline and joined to the sea soon after the end of the last Ice Age, some 15,000 years ago, according to John Rees of the British Geological Survey in Nottingham. Analysis of lake sediments shows that the weight of ice had depressed Scotland to the point that, given the rapidly rising sea levels of the day, marine waters extended further inland than they do today. Scotland has gradually risen over the past 15,000 years, rather like a mattress bounces back when you get off it, severing the connection with the sea. That left a number of so-called meromictic lakes, which started out as a sea loch or fjord that was marine and was subsequently isolated from the sea due to the rebound from the Ice Age. These lakes, notably Loch Lomond, still had salt water at their bottoms (preserved due to physical depth, density contrast, and lack of flushing by fresh water). Unfortunately, while no examples remain today due to high rainfall, some lochs are brackish, and perhaps the Hogwarts squid has adapted to the lower salinity.

There are also examples of sea lochs that, at first glance, can be mistaken for isolated lakes, according to Graham Shimmield of the Dunstaffnage Marine Laboratory in Oban, Argyll, on the west coast of Scotland. One example, called Loch Etive, is next to his laboratory. The Falls of Lora and the Bonawe narrows are the two 'sills' separating the main loch from the sea. 'The maximum depth of Loch Etive is over 140 metres – deeper than the continental shelf to the west of Scotland, and the North Sea – a perfect hiding place,' says Shimmield. Now his team is looking for other examples, notably Loch Morar, and some of the Outer Isles lochs on the Uists. All of these are on the west coast, where warmer waters are provided by the Gulf Stream, corresponding with the mention of warmer waters that the creature likes to bask in. One day, these clues may reveal the true location of Hogwarts and its lake.

Blast-ended Skrewts

A fascinating piece of natural firepower can be found in these highly dangerous crosses between a manticore (a mythological beast with the head of a man, body of a lion, and scorpion's tail) and a fire crab (a bejewelled tortoiselike creature equipped with a flamethrower). When Harry first encountered them, the Skrewts looked like deformed, shell-less lobsters with legs in odd places and no visible heads that could deliver the occasional blast. By the time he competed in the Triwizard Tournament, one specimen had grown to ten feet long and, with its sting, looked like a giant scorpion.

As we will see, the techniques of genetic engineering offer many opportunities to craft a Skrewt, but there is one feature that, surprisingly, can already be found in a real-world creature. Nature has found a clever way to put the blast into the ends of the Skrewts.

The Skrewt probably uses a rocket propulsion system of the kind that sent German V1 'doodlebug' robot bombs to southern England during the Second World War. Though the doodlebug was the very first operational cruise missile, the propulsion method was not in fact first developed by the Nazis but by the bombardier beetle, an insect that can produce a spray of chemical irritant at a temperature of 100 degrees C about 500 to 1,000 times per second.

The bombardier's secret is a pulsed rocket mechanism – little explosions that generate spray – very similar to that of the V1, according to research by Thomas Eisner of Cornell University. 'What is interesting about the V1 and the bombardier beetle is that they both have a system where the valve that provides fuel to the reaction chamber oscillates passively – the first explosion closes the valve that feeds fluid into the explosion chamber,' says Eisner. After the first explosion, reservoirs of chemicals in the beetle's abdomen are drawn into a reaction chamber by a drop in pressure. There an explosive reaction occurs to eject yet another hot mist of irritant chemicals called benzoquinones. The beetles can spray the boiling, toxic fluid from the tip of their abdomens in virtually any direction by moving the tip and using a pair of shieldlike deflectors. To imagine that a young Skrewt used similar technology to propel itself forward a few inches with a small *phut* does not seem a great stretch of the imagination.

My confidence that the Skrewt is using similar technology is bolstered by the knowledge that these blasts have been a feature of wildlife for a very long time: evolutionary evidence points to the bombardier beetles having evolved before the break-up 150 million years ago of Gondwanaland, the supercontinent made up of the current continents in the Southern Hemisphere.

Poisonous orange snails

Although these creatures receive only a brief mention in the Harry Potter books, they must surely be a reference to famously toxic carnivorous marine creatures called cone shell snails, according to a local Hogwarts gastropod expert, Les Noble of Aberdeen University. There are over 500 species of cone shells worldwide, ranging in length from 1 centimetre to 15 centimetres. Because of their diverse beauty, they have been collected for more than three centuries. But the snails can also be deadly, producing venom in a special duct and poisoning their prey by injecting a venom rich in 'conotoxins' with a hollow harpoon-like tooth. Fish, worms or molluscs are injected with the lethal cocktail during the snails' nightly hunting missions.

Some fish-eating species of the snails are so toxic that they can even kill people. Over thirty cases have been recorded in the scientific literature. In 1936, *The Medical Journal of Australia* described how a healthy young man had a fatal encounter with the fish hunter *Conus geographus* while visiting Hayman Island on a pleasure cruise with his mother and friends. He died five hours after being stung. The symptoms resembled those of poisoning by curare, including pain, numbness, vomiting, dizziness and respiratory paralysis.

The presence of orange snails at Hogwarts could mean that dark work is afoot. Indeed, one can easily imagine a shaman from ancient times finding many uses for these beautiful creatures. Bruce Livett, a cone snail expert at the University of Melbourne, believes several types of cone snail fit the description in Harry Potter. For instance, there is the unusually large – 60 millimetre – fully orange specimen of *Conus regius citrinus* found by the collector David Touitou in Martinique in the French Antilles. Then there is an orange speckled *Conus spurius* Gmelin that lives in Colombia, or even *Conus capitaneus* which can be seen in a floral still-life by Balthasar van der Ast (c. 1593–1657) at the Rijksmuseum in the Netherlands. But it seems unlikely that cone snails would be put to decorative use at Hogwarts. Another, perhaps more promising, candidate for the poisonous orange snail is *Conus textile suzannae*, found in Kenya, which has been suspected of inflicting pain on humans.

Cone shell snails are not all bad. Venoms act by interfering with molecular signalling pathways in the body and this mechanism can be put to good use. Snail venoms typically carry not just the four or five components typical of a spider or snake venom but as many as 200, making them a valuable source of potential new pharmaceuticals. 'The great diversity of cone shells, each with a unique cocktail of toxins, offers great possibilities for molecular prospecting for novel drugs from the sea,' says Livett.

Tiny proteins or peptides have been isolated from the venom of fifty species of Great Barrier Reef snails by Paul Alewood and colleagues at the University of Queensland, revealing several with great potential as highly selective pain-killing drugs. There is also evidence that these analgesics could be derived from a poisonous orange variety, such as *Conus consors*, a fish-hunting cone snail. A team at the Institut Fédératif de Neurobiologie in Gif-sur-Yvette, France, discovered that its poison contains a novel toxin – unmagically named omega-conotoxin CnVIIA – which blocks the channels that allow charged calcium atoms (calcium ions) to pass into nerve cells. 'Other cone shell toxins that block these N-type calcium channels have been shown to be potential blockers of pain, in particular of neuropathic pain, the type of pain that is resistant to morphine treatment,' says Livett.

One such cone shell toxin, discovered by Baldomero 'Toto' Olivera and colleagues from Salt Lake City, has already undergone clinical trials for use in addressing neuropathic pain where other treatments have failed. The analgesic, called Ziconotide, is based on the conotoxin of the marine cone snail, *Conus magus*, which is usually yellow or brown on a white background. Intriguingly, however, Livett points out that there is evidence for an orange-marked version in one of the cone snail bibles, *Manual of the Living Conidae* by Dieter Rockel, Werner Korn and Alan Kohn. Thus it seems that the orange poisonous snails of Hogwarts could have a more benign use, helping to ease the persistent pain of an injury sustained in Quidditch or caused by a spell that went disastrously wrong. The snails probably feature in one of the recipes in Potter's copy of *Magical Drafts and Potions*.

Natural wonders

The squid, snail and Skrewt underline how, even without magic, the Muggle world has many exotic and intriguing creatures that sound like they ought to inhabit the enchanted world of Hogwarts. Imagine, for example, an infection that murders males, triggers virgin births and turns boys into girls. Feminist fantasy? Misandrist propaganda? Science fiction? No. Welcome to the bizarre world of wolbachia, the gender bender, widowmaker and slayer of males, a bacterium that toys with the sex lives of an estimated 20 per cent of all insect species, from wasps and butterflies to ladybirds.

There are other extremely odd creatures in the Muggle world. Some exert the equivalent of the Imperius Curse, one of the unforgivable kind that puts the victim under the control of whoever casts the spell. *Dicrocoelium dendriticum*, the lancet liver fluke parasite of cattle, infects ants and alters their behaviour. When the temperature drops as evening approaches, infected ants climb atop blades of grass and clamp on to the leaves with their mandibles. There, the ants wait to be eaten by browsing cattle. In that way, the parasite's extraordinary life cycle is completed.

Love potions even have an odd Muggle counterpart. There is a 'love-bug' organism that causes rats to abandon their innate aversion to cats. *Toxoplasma gondii*, an intracellular protozoan, infects the rodent's brain, inducing an effect similar to that of Prozac, so it becomes less fearful of cats. Once eaten by the cat, the parasite is successfully transmitted to its definitive host, which ensures the completion of *T. gondii*'s life cycle.

Nor does the transformation of an animagus, or cross-species switches, seem so magical when compared with the bizarre lifestyle of a creature discovered by Peter Funch and Reinhardt Møbjerg Kristensen of the University of Copenhagen. It was adhering to the mouthparts of a Norway lobster where it swept up errant food particles, like a living napkin. The creature is so unique – or peculiar, depending on your point of view – that the biologists gave it not just its own species name, or its own genus or family, but have moved it way up the classification scale and declared that it is an entirely new phylum. This is taxonomy's jackpot, for there are only thirty-five or so known phyla, the second-highest ranking in the classification scheme, just below the uber-designation of 'kingdom' that separates animals from plants from fungi.

The name of the new phylum is Cycliophora, Greek for 'carrying a small wheel', a reference to the creature's very striking circular mouth, located next to its anus. The somewhat magical species name they bestowed on their find is *Symbion pandora*. The first part of the name refers to the animal's status as a symbiont, a creature that co-exists with another (the lobster), and the second to its intricate life cycle. The convoluted cycle comes in two parts, an asexual one and a sexual one, each of which consists of several different stages. As Funch says, 'Investigating the life cycle was like opening Pandora's box.' Many of the denizens of Hogwarts are magical, but can they really compare with *Symbion pandora*? Apart from the occasional indirect reference to microscopic life, notably viruses that cause colds and influenza, or the wand-munching Chizpurfle parasite, we can't yet rule out the possibility that a lobster lurking at the bottom of Hogwarts lake is infested with something even more peculiar than *Symbion pandora*. Indeed, it could be teeming with magical microscopic life. And, even if such odd creatures do not exist, it is increasingly possible that we could manufacture them to order, as we are about to see.

CHAPTER 6

Magizoology

Cerberus, the cruel, misshapen monster, there
Bays in his triple gullet and doglike growls
Over the wallowing shades; his eyeballs glare
A bloodshot crimson, and his bearded jowls
Are greasy and black; pot-bellied, talon-heeled,
He clutches and flays and rips and rends the souls.

Dante, *The Divine Comedy*

Many of the creatures that Harry Potter encounters could have been the result of the efforts of a crazed genetic engineer. Think of the climax of Harry's first wizard outing – when he walks through a curtain of black flames to encounter Quirrell, a supposedly shy and nervous young teacher. Quirrell unwraps his turban and Harry is horrified to see a white snakelike face. The Dark Lord – Voldemort – had somehow taken up residence in the back of Quirrell's head, a grotesque sight that would haunt Harry's dreams for weeks to come.

Think of the gargantuan girth of Rubeus Hagrid, the beetle-eyed Keeper of the Keys and Grounds at Hogwarts. Is his giant size in some way related to the action of Swelling Solution, which makes any part of the anatomy expand on contact? Was Hagrid once blown up in the same way that Aunt Marge once inflated into a piggy-eyed balloon? This magic seems powerful: it may also have created the giant spiders in the Forbidden Forest, such as Aragog, or the Nundu, a gigantic leopard-like creature. Perhaps Shrinking Potions caused equal-but-opposite effects.

Hogwarts is populated by a vast and fantastic array of life. Magical plants sprout in the greenhouses. There is Fluffy, the three-headed monstrous dog with yellow fangs (though how Fluffy has evaded the Committee for the Disposal of Dangerous Creatures, goodness knows). The Magical Menagerie contains custard-coloured furballs, double-ended newts, huge purple toads and much more besides that squeaks, howls or hisses. And, of course, there is Dobby the obsequious house elf, a little creature with bat-like ears and pop-out eyes the size of tennis balls.

Talking of Dobby, those leathery-looking gnomes, which resemble muddy potatoes on legs, seem somehow related to him. Measuring just ten inches high, with horny feet and large, knobbly bald heads, are they in turn cousins of Red Caps, those goblins who have a thirst for bloodshed? How do they relate to tree-loving Bowtruckles, those fairy-like Doxies, or elfish Erklings who have a taste for Muggle children?

A comparison of their genetic recipes, or genomes, could untangle the roots of their family trees. Fortunately for our quest to understand the strange creatures that inhabit Harry's world, we can benefit from the many international programmes under way to read the genomes of a host of creatures, from human to worm, fly and bug. These projects are providing plenty of insight and understanding for any wannabe wizard gene tinkerer.

Only recently, scientists also deciphered the entire genetic recipe of a humble weed called *Arabidopsis* which is similar to the recipes used by other plants. There is much evidence that these instructions have been tampered with at Hogwarts. Consider the 'umbrella-size flowers' dangling from the ceiling of Greenhouse Three. They may have been created using the same genetic growth recipe as the cabbage-size flowers and the boulder-size pumpkins that Hagrid raises in the small vegetable patch behind his house. Could the *Arabidopsis* code also hold the secret of the dynamic abilities of the ancient Whomping Willow, which has developed the means to attack people with its branches? Perhaps the same magic is at work in the bucket of jumping toadstools.

Our understanding of why things are the particular shape they are – and our ability to alter that shape – is the culmination of an effort that dates back millennia. Hippocrates, in the fifth century BC, couched the matter in terms of fire and humidity, wetness and solidification. Aristotle posed one of the key questions a century later: Do all the parts of the developing embryo begin together, or do they appear in succession? We now know that the latter is the case, and we also understand the molecular processes at work, which offer profound opportunities to understand – and thus change – a developing body plan.

Imagine you had a basic set of building blocks to create any body, just as Lego bricks can be used to build all sorts of toys. Then it would be easy to rustle up a fantastic creature, whether a unicorn, centaur or giant spider. It sounds like fantasy, but over the past few decades we have discovered that Nature uses this very approach. Biology has seen a revolution in understanding of development as a result of studies of a range of creatures, many of whom would feel at home in Hogwarts, notably mice, frogs, flies and the hydra, a creature that can regrow lost heads, or even a whole body, from a tiny fragment of the original.

The Lego of life

When sperm meets egg it triggers a cascade of chemical reactions, hundreds of thousands of which follow one other, overlapping and crossing in a network of unbelievable complexity. A fertilised egg carries no preformed tiny homunculus (or manikin) of the body-to-be. Instead it carries genes, instructions held on DNA molecules, which tell it everything (almost) it needs to know about life. Somehow the genes must respond to local geography in the bud of a limb or the middle of the foetus to sculpt a body.

When a human egg develops into a body, the molecular mechanisms that generate the 'body plan' are very similar to those used by the mouse, fruit fly and other so-called primitive creatures. Every living creature, from insects to people – even, I suspect, wizards and witches – is built with the help of very similar assembly kits. The components are the genes that we inherit from our parents.

Take a close look at any creature and you can see basic building blocks, whether they take the form of the segmented bodies of worms, vertebrae of animals with backbones, or the head-thorax-abdomen construction of insects. There are even more peculiar correspondences between the building blocks of different families. At the National Institute for Medical Research in north London, a team led by David Wilkinson and Robb Krumlauf found similarities in the way a mouse develops segments in its brain and the way fruit flies form segments of their bodies.

Segmentation, like bricks of Lego, is a simple way to build something complicated like a living organism. In a worm, all the individual segments look alike, but more complex creatures reveal differences: insects, for example, develop wings and legs on different segments. In even more complex structures, from men to mice, this segment programming is used as a basic unit on which rococo variations and designs can be made.

In humans, although segmentation is an evolutionary relic, it is used to develop the body plan. In particular, tiny segments corresponding to gills can be seen in the early embryo similar to those in fish embryos. While in fish these segments do develop into gills, in men or mice these 'gill arches' develop to form part of the face, including the lower jaw and neck. This is very suggestive. When Harry ate Gillyweed to grow gills so he could breathe under water, did he reactivate this genetic programming?

The development of the hind part of the brain (constituents called rhombomere segments) has started even before the gill arches. In this way, one can build a head. So when Ron Weasley writes to Harry about Egyptian curses that made Muggles sprout an extra head, or Dark wizards crave a three-headed Runespoor serpent, one can see how this extra capital could be achieved in broad outline: all one has to do is duplicate the fundamental genetic programming.

How to plan a body

The study of genetics of development was born when scientists saw how monsters resulted when the process of segmentation went awry. Occasionally, an insect appears with the 'wrong segments': a leg where there should be an antenna, for example. As long ago as 1894, the British scientist William Bateson dubbed these 'homeotic mutations', where 'homeosis' refers to changing one part of the body into the likeness of another. Bateson suggested they might throw light on the secrets of evolution and embryonic development, an instinct that proved correct.

A quarter of a century later, the first of a strange class of fruit fly mutations was found. An insect appeared with two copies of the segment that carries its wings. These flies, which sported two sets of wings instead of the usual one, triggered an avalanche of discoveries that represents a revolution in our understanding of the molecules that mark out the body plan.

Using the fruit fly, one of molecular biologists' favourite

experimental creatures (they are easy to study, their genes are easy to manipulate and, er, they breed like flies), scientists found that a tiny gene sequence, called a homeobox, appeared in many of the genes that, when mutated, cause mutated bodies. All body plan genes contain the homeobox so by hunting down the homeobox in normal cells, scientists could uncover the genes involved in the molecular mechanism of development or pattern formation. When a fly embryo is four hours old, for instance, a gene called Engrailed gets switched on. This particular homeobox gene helps the body divide up into segments.

When the first homeobox genes were isolated in the fly they turned out to lie in clusters along the genetic code. Similar clusters have since been found in the DNA of other animals: worms, mice, frogs, chickens and humans. Not only do these clusters of genes live together, they work together. Each cluster forms a team; the team members tell the host cell where it is in the body, and thus what to become.

Genes containing homeoboxes are crucial for shaping bodies because they produce proteins that are used to control how the genetic recipe is read and interpreted. Thus, in arthropods, a group that includes insects, spiders, crabs and centipedes, the first genes in the cluster influence the development of the head and associated structures, those in the middle influence development of the legs and wings on appropriate body segments, and so on. In humans, the cluster has been repeatedly duplicated to form four clusters, all slightly different, with a total of thirty-eight genes. One role of the homeobox genes is to control limb development, dictating whether a group of cells becomes a thumb, a little finger, or anything in between. In rough terms, clusters become larger with increasing body plan complexity. This might suggest that the basic recipe varies. However, the development of body plans in all animals is controlled by a remarkably small number of genes.

How Harry Potter got his shape

Equally extraordinary is that each and every cell in a developing embryo has exactly the same genetic information in its DNA. So how do the cells in Hagrid's head, or Harry's for that matter, end up different from those in his heart? For decades there has been a debate about whether early cells in the embryo are preprogrammed to become, say, heart cells, or whether they are told to change by their sister cells. Scientists now believe that the environment of each cell in a developing embryo is all-important; a heart cell is the product of nurture, not nature. Move cells around in an early embryo and they will change identity, depending on their local circumstances.

Patterns are laid down as one cell influences the identity of its neighbour, passing the word – 'We are going to be a heart' – like a rumour spreading through a crowd. But there are also signals that act rather like a director on a Hollywood film set: 'You guys will be a heart!' These are called morphogens, signalling chemicals that organise a matrix of surrounding cells into patterns, and have been studied by Sir John Gurdon, based at Cambridge University, who has done much of his pioneering work in the field on frogs.

Whether they are local cellular signals or morphogens, these chemicals turn on homeobox genes which seal the fate of a cell, turning it into a heart, liver, skin cell or whatever. Regulatory (homeobox) genes that are active early in the process of development help determine which end of the embryo becomes the head and which the tail, which part becomes the back and which the belly. These genes also set up the basic tissue types, turning an embryonic tissue called ectoderm into skin and brain, for example. Genes that are active later in the cascade help block out distinctive segments within the body – differentiating, say, a head from an abdomen. Later still in the cascade, genes influence the growth of appendages such as limbs, until the most refined morphological details have been achieved. The body plan is thus a pattern in time as well as space.

When a homeobox gene is faulty, however, peculiar things happen and the monsters that result can reveal a lot about what a particular gene does in the body. In fruit flies, for example, a gene called Antennapedia normally turns on only in the cells that make up the thorax to tell them that protuberances on the embryo's surface should develop as legs. When it goes wrong and is turned on in the head region, the insect will sprout a pair of legs where its antennae should be. (This sounds rather like that disturbing reference in *Moste Potente Potions*, when a witch sprouted several extra pairs of arms from her head.)

Fruit fly groups around the world have assembled an array of mutations of mythic proportions. For example, Bicaudal has no head, not much of a body, and two anuses. Other mutant flies are hairy, bald, eyeless, tumour-ridden, all front and no back, all back and no front. Then there is our old friend Bithorax which has two sets of wings rather than one.

Now we have at least an idea of how to create the winged palominos that dragged along the Beauxbatons' carriage and many other creatures besides. As Matthew Freeman of the Laboratory of Molecular Biology in Cambridge notes, what is truly amazing is that the cells in all creatures tell each other what to do in the same language. 'This comes in handy if you want to make fantastic creatures. And you can see how Rita Skeeter can become a beetle, Dudley Dursley can sprout a curly pig's tail, a Quintaped can end up with five hairy legs, and that it might also be possible to mix and match features to make chimeras.'

Nor does the idea of transformation sound so unlikely, whether it involves the Greek myth of the enchantress Circe who turned men into pigs, the Streeler snail that changes colour on an hourly basis (according to Scamander), or the Animagus, a human who can turn into an animal at will. All living things seem to be variants on a basic genetic theme.

Let's look at how to create a magical menagerie. None of what follows would be ethical in the Muggle world. After all, gene tinkering is at a primitive stage of development, the public are wary of genetic manipulation and have been for some time (think of *Frankenstein* and *The Island of Dr Moreau*, for example) and the vast majority of scientists have many more scruples than they are given credit for. Making magical creatures is not their goal.

In the wizarding world, Newt Scamander was responsible for the Ban on Experimental Breeding, passed in 1965, which prevented the creation of new monsters in Britain. In the Muggle world, the UK Advisory Committee on Releases to the Environment seems to have adopted his tough stance. The Committee would never pass an application to release Fluffy, fearful of what would follow, let alone allow scientists to breed the three-headed slavering beast in the first place.

The idea of creating a menagerie of mythical beasts would also be anathema to the Agriculture and Environment Biotechnology Commission. The commission, which takes a more strategic view of genetic wizardry in the UK, would not approve of any attempt to create the Hogwarts bestiary, according to one of its members (who did not want to be named, lest he seemed to be encouraging magizoology). Fortunately, as J.K. Rowling makes clear, Clause 73 of the International Code of Wizarding Secrecy demands the concealment of the magical bestiary of Harry Potter's world from Muggles so as not to increase the level of hysteria about genetic modification.

The following fantastic creatures, no doubt created before the 1965 ban, can all be found in safe habitats, protected by Muggle-repelling charms. I hope that the following discussion will provide some insights into how to create them. But any attempt to publicise details about them will, of course, be met with a swift response from the Office of Misinformation with a Disillusionment or Memory Charm.

Dobby the house elf

Dobby had great green tennis-ball eyes, a pencil-like nose, batty ears and long fingers and feet. He was a house elf who served the Malfoy family and seems to be related to Muggle brownies, the home spirits who, in Scottish superstition, are supposed to do jobs for families and, by one account written in 1584, 'would chafe exceedingly, if the maid or goodwife of the house, having compassion of his nakedness, laid anie cloethes for him, besides his messe of white bread and milke, which was his standing fee. For in that case he saith; What haue we here? . . . here will I neuer more tread nor stampen.' (This, perhaps, explains why Dobby left the Malfoys when presented with a sock.)

To create a Dobby, or even a female of the species such as Winky for that matter, it would be easiest to start with a similar creature to keep the amount of gene manipulation to a comfortable minimum. One good candidate would be found in Madagascar: a goggle-eyed, bat-eared little ball of black, wiry fur known as an aye-aye, an endangered species and one of the most peculiar of all primates.

This nocturnal animal has been compared to a gremlin or gargoyle and resembles a cross between a bat, a beaver and a racoon. So queer is the aye-aye that when it was discovered in the eighteenth century, zoologists first classified it as a squirrel or even a kangaroo. When they finally realised it was a primate, they marked the discovery with a scientific name, *Daubentonia madagascarensis*, that made it the sole member of its genus. Perhaps the wizarding world shortened *Daubentonia* to give us the name of Dobby.

Dobby seems somewhat garrulous when he first meets Harry, and in this respect the powerful jaws of the aye-aye are an ideal model: they can gnaw through a concrete block wall. While Dobby had a long, thin nose, and looked somewhat like an ugly doll, the aye-aye is a bit hairy and ratlike, so we need a selection of human genes to make him look a little more like us, and raise his IQ. Perhaps we also need some extra machinery to endow him with the bigger eyes typical of his relatives, the western tarsier, South African galago or the slender loris.

The hands of the aye-aye will need some work: the creatures possess an elongated middle finger that they use as an extremely delicate probe to hunt for tasty insects in wooden nooks and crannies. The other fingers will have to be extended so that the middle finger does not stand out so much, which would be to their benefit, for it is less likely to be noticed by superstitious Malagasy, some of whom believe that if an aye-aye points its finger at them, they are doomed to die. They persecute these 'witch creatures' and kill the unfortunate animals on sight.

How to make Hagrid

Many oversize creatures wander around Harry Potter's world. Take for example, Hagrid, whose mother was the giantess Fridwulfa (that is, if you believe everything you read in the *Daily Prophet*). Any consideration of stature leads us naturally to Sir Paul Nurse, a diminutive geneticist who, inspired by abnormally small yeast cells called wee mutants, went on to win the Nobel Prize. He has also been known to wear Harry Potter's pointed hat in King's Cross Station to promote his charity, Cancer Research UK. The wee mutants that he studied paved the way to understanding cell division, one of the most basic processes that builds and maintains a body, and also goes awry in cancer. Sir Paul points out that you can boost the size of any creature in one of two ways. One is to increase cell size, rather like building a house out of bigger bricks. 'If you delay the cell cycle with variations of cell cycle mutants, cells grow larger and larger. They are opposite to wee,' he explains. Alternatively, one could boost the number of an organism's cells, or increase their number and size.

When Hagrid grows, cells divide in his body. An original cell divides into two daughter cells, each of which has half the mass of the mother cell. Those daughters then need to grow up and reach a critical size before they can divide again. Among scientists, there has been a debate about how cell growth (increase in cellular mass – through the manufacture of protein) and cell division (the generation of more cells) are coordinated to boost body size.

Studies of experimental organisms have shown that there is no single answer to this question. Mammals may do it differently from yeast or fruit flies, for example. But it is clear that cell size does influence the onset of cell division. Not surprisingly, eating in turn influences size. In fruit flies, size is controlled by a metabolic process called the insulin pathway, according to Ernst Hafen of the University of Zurich. Less insulin signalling makes smaller flies, while more signalling makes them bigger. As insulin is a well-known regulator of metabolism and is itself controlled by how much we (or a fruit fly) eat, it seems to be a key player in making sure that fruit flies grow when they are well fed but stay small (though perfectly formed) when there's not much food around.

The varying strategies used to control the size of different animals is nicely illustrated by a gene that has been studied by hundreds of scientists for many years. Called Myc, it was first identified as being involved in cancer but more recently has been investigated for its role in regulating size. Reducing the amount of Myc in flies makes them smaller. When Peter Gallant, also in Zurich, examined them, he found that all the cells had shrunk, so everything about them was tinier. A similar experiment was carried out in mice by Andreas Trumpp of the Swiss Institute for Experimental Cancer Research in Epalinges and colleagues from California. Like the flies, the mice were smaller but all their cells remained normal size though there were now fewer of them.

In mice, the Myc gene seems to control cell division, not size, itself. It is this important role that probably also underlies Myc's function in cancer, which is primarily a disease in which cell division becomes uncontrolled. Quite why the same gene should control body size in such fundamentally different ways remains a puzzle, but these experiments do highlight the importance of Myc and suggest that it could be the target of magical powers. For example, the gene may have been affected in some way when Dudley Dursley ate one of Fred Weasley's Ton-Tongue toffees. Did Myc make his tongue look like a slimy purple python?

Muggles have already tinkered with genes to make over-sized creatures. Laboratory mice which have been genetically altered to produce human growth hormone grow to be 25 to 30 per cent larger than normal mice – with much of that size difference being due to bigger bones, according to a study done at the University of Michigan by Steven Goldstein. However, as ever with a science in its infancy, this approach may have problems. In this case, Goldstein has observed bone fragility. This is reminiscent of a classic cautionary tale about genetic alteration, concerning the Beltsville pig – in fact a number of pigs – which were genetically engineered more than a decade ago to produce human growth hormone by Vernon Pursel's team at the US Department of Agriculture in Beltsville, Maryland. Although the team hoped the pigs would grow faster, most in fact suffered crippling arthritis because the hormone was produced in the wrong tissues. The experiment was immediately halted, and an important lesson had been learned by the genetic engineers. The work, which underlined the hubris of some of the early optimistic claims made for the technology, was seized on by campaigners against genetic modification as evidence that the technology was dangerous. However, what it really emphasised was that the new biology has yet to find its feet.

The Shrinking Solution

At first glance, this spellbinding solution should have the opposite effect of the growth formula. But it becomes clear from Harry's experience in his double Potions lesson that its effect is more subtle and interesting than mere shrinking. Professor Snape explained that for the solution to work, it should not only cut the size of a toad but turn it back into a tadpole. Somehow the solution reversed the developmental programming in Trevor the toad, the experimental subject provided by Neville Longbottom.

An army of scientists around the world would dearly like to be able to carry out de-differentiation – reversal of the specialisation of tissues – as easily as can be achieved with Shrinking Solution. If only they had Snape's recipe. At the moment, the one surefire way we have to turn an adult cell into an embryonic one is by cloning, a possibility that was initially raised by Sir John Gurdon working in Cambridge in the 1960s using gut cells from *Xenopus*, the African clawed frog. In 1997 Ian Wilmut at the Roslin Institute and PPL Therapeutics, both near Edinburgh, announced the world's first clone of an adult mammal: a Finn Dorset sheep called Dolly. She was created using what is now called the nuclear transfer method, where an egg is 'reprogrammed' with the genes of an adult cell – in Dolly's case a mammary cell.

The really clever part of the technique overcomes one key obstacle in cloning from adult cells. Although adult cells contain an organism's entire genetic code, they have become specialised. In other words, they use only a small proportion of their genes, those relevant to functioning as a heart cell, skin cell or whatever. The remaining genes lie dormant. The research by Wilmut and his colleagues demonstrates the ability to create de-differentiated individual cells from the tissue of an adult.

Now scientists are studying how to achieve this without the need for eggs and cloning. Martin Raff of University College London found that extracellular signals can cause at least some cells to de-differentiate by turning on some genes and turning off others. It is still unclear whether the cells simply retrace, in reverse, all the steps they originally followed in differentiating and then take a new differentiation pathway, or whether they simply take a short cut to the new differentiated state. But to turn a frog into a tadpole, you would not only have to turn back the developmental clock but also find a way to revive cells that died off during development. 'If you could de-differentiate most of the toad cells to their original tadpole state and reform the cells that died during metamorphosis, you could, in principle, turn a toad into a tadpole,' Raff says.

This work on winding back the developmental clock is being pursued with vigour across the planet because scientists want to use the method to create embryonic stem cells, the grandmother cells of all other cell types. As will be seen in the chapter on the philosopher's stone, these cells can make any tissue type, from bone to brain, to repair an aged or damaged body.

The ultimate stem cell is the fertilised egg. As it divides, its progeny become even more suited to a particular task or organ, until there is a wide repertoire of possible cell types, each of which carries out a specific job, whether in brain, muscle or gut. En route to adulthood, an individual develops even more specialised stem cells. For example, there are blood stem cells, from which all the blood's red and white cells arise, stem cells that proliferate to form and renew the skin, and stem cells that form the dazzling complexity of the brain.

While at Cambridge University, Martin Evans became the first to isolate mouse embryonic stem cells. Huge numbers of the cells could be grown over long periods in the laboratory, providing new opportunities for genetic experiments. Then James Thompson and colleagues at the Wisconsin Regional Primate Center in Madison, Wisconsin, became the first to grow stem cells from an early primate embryo, with a technically challenging procedure in which the cells were bathed in special factors and grown on a bed of sterilised rat cells. Worldwide, scientists are now studying how to use human stem cells in treatments of a number of conditions. Stem cells could be used, for example, to fix a damaged heart, or create dopamine-producing nerve cells for the brain of a Parkinson's patient. No doubt, the Hogwarts matron already relies on all kinds of stem cell sorcery in tending to her ailing charges.

Aintegumenta! and Whomping Willows

Aintegumenta may very well be what a wizard or witch bellows to create those umbrella-sized flowers and boulder-sized pumpkins that grow in the magical gardens and green-houses of Hogwarts. Aintegumenta, nicknamed Ant, is a spell that is also often muttered by plant scientists when trying to understand why plants grow to be the size that they are.

Ant is a gene that gets its name from the word 'integument', meaning an outer protective coating – for instance of a seed, or rind. The first hint that the gene plays a role in plants' organ growth came from David Smyth's lab at Monash University in Australia, where the molecular identity of the gene was determined, along with Robert Fischer at the University of California, Berkeley. Based on this work, Beth Krizek at the University of South Carolina discovered that Ant appears to control the growth of floral organs (sepals, petals, stamens and carpels) and leaves in studies using the geneticist's favorite plant, *Arabidopsis*. Then the gene was found to regulate growth and cell numbers during the development of both leaves and flowers by Yukiko Mizukami who was working with Fischer at the University of California, Berkeley.

They reasoned that because flowers, petals and leaves always tend to grow to the same size in a given species, they must be under some kind of genetic control. In other words, each plant has an intrinsic size, a gene-based programme that says grow so much and no more. Ant turned out to be part of that control mechanism and may work by limiting the number of cells in each part of a plant rather than the size of each cell. The gene is thought to keep cells in a state in which they can continue to grow and proliferate. Turn Ant off, and growth halts. Turn it on and keep it on, and you may end up with a shed-sized pumpkin.

One of the more startling examples of genetically modified plants in Harry Potter is the Whomping Willow, which can move swiftly and thus has the means to attack people and crashed cars with its boughs and branches. All plants have some power of movement, which may be as simple as growing a little bigger or unfurling a flower in the sunshine. But carnivorous plants, for example, show that the movement of a plant can be fast, despite its lack of muscle, providing clues as to how a willow could whomp. Indeed, similar magic may have been at work to flip fungi in that bucket of jumping toadstools.

Venus's fly-traps harness changes in water pressure to snap shut. When the trap is activated (for instance, when a fly triggers hairs on the leaves), the cells on the inside walls of the trap transfer water to the outside walls. This drop of pressure inside the walls and the rise outside them snaps the leaf closed. Another kind of motion in carnivorous plants is powered by cell growth. The tentacles of sundews, for example, bend towards prey because the cells on one side of the tentacles outgrow the cells on the other side.

In terms of sheer bulk, the largest carnivorous plants are

Nepenthes vines, which can grow tens of metres in length. Plants in this genus also have traps that have evolved to capture some of the larger prey, including creatures the size of frogs, and even birds or rats. On a larger scale, this could be the model for a hydraulic technology capable of packing a whomp into a willow, or indeed could make a devil's snare creeper more gripping. When Hermione used her wand to set Ron and Harry free from its tenacious hold, did its spurt of bluebell flames disrupt the snare's hydraulics?

The Hinkypunk

This strange will-o'-the-wisp stands on a single leg and consists of evanescent whorls of smoke. The Hinkypunk supposedly hops around with a lantern, which the creature uses to lure anyone walking nearby to fall in a ditch, stumble into a bog, and so on. While this creature may sound somewhat preposterous, genetic science has come up with a much more plausible explanation for its existence: the Hinkypunk could be a swarm of bacteria in a puddle of bog water. When this colony grows to a certain size, it can give off a flash of light to confuse any innocent passer-by.

Understanding of how the Hinkypunk can be so misleading has come from a research project that started by asking esoteric questions about certain bacteria that are ubiquitous in the world's oceans. Called *Vibrio harveyi* and *Vibrio fischeri*, both emit a blue glow. Bonnie Bassler of Princeton University discovered that to become flashy these bacteria carry out two forms of what is called quorum sensing, a term coined by Pete Greenberg of the University of Iowa (whose son, Ted, is a Harry Potter fan). Each bug makes a small quantity of a signalling molecule that builds up in concentration as a population grows. These concentrations of signalling molecule are used to sense whether the bacteria are part of a dense or sparse population. When there is enough chemical present, the bacteria adjust to their crowded environment, and for *V. fischeri* and *V. harveyi*, the response is to emit a blue glow. In fact, *V. harveyi* has two quorum-sensing systems, either of which can trigger the gleam: one system tells the bacteria how many of its own species are in the area; the other indicates how many other types of bacteria are around. This work has an intriguing implication: even lowly bacteria 'talk' to each other, just like the cells in our bodies.

Now we can see how the Hinkypunk could work. Colonies of these bacteria could thrive in a boggy puddle. When they reach a certain size, they could give off a blue glow to confuse and confound an observer. A circular puddle, when viewed from a distance, of course, would create the illusion of something long and thin. It does not seem too far-fetched to suggest that it could, in the dead of night, appear to be a one-legged glowing creature, particularly when it is glimpsed by a stranger with a fertile imagination who has not encountered a Hinkypunk before.

The bioluminescent bacteria that power the Hinkypunk are in fact no strangers to deception. They are present in seawater and colonise a special light organ in the body cavity of a secretive Hawaiian squid, *Euprymna scolopes.* This fascinating symbiotic relationship was explored at the University of Hawaii by Edward Ruby and Margaret McFall-Ngai, along with Karen Visick at Loyola University in Chicago. As well as deploying the conventional defence against predators, by releasing a squirt of ink, the squid also harnesses the bacteria in a cunning example of natural stealth technology. While feeding at night, it uses the bugs to emit light downwards and matches it to the intensity of moonlight overhead, rather like the enchanted ceiling of the Great Hall at Hogwarts. That way, it confuses any predators. The bioluminescence allows the animal to blend in with the ambient light levels, which results in the absence of a shadow, confusing any predators nearby.

Nifflers

These fluffy black creatures with long snouts and spade-like front paws sound a lot like moles, only Nifflers like to hunt for bright objects – which suggests that, unlike moles, they have good eyesight. One possibility is that they have had light sensors or eyes added. But how do you build an eye? The puzzle of how an organ as exquisitely complex as an eye could evolve at all was so problematic that Darwin felt compelled to discuss it at length in his book *On the Origin of Species* in a chapter entitled 'Difficulties of the Theory'

Since Darwin wrestled with how to construct an eye, much progress has been made in understanding the process by which it comes about. Today, we realise that to create a Niffler, one simply has to refer to eye-opening experiments conducted a few years ago by Walter Gehring of the University of Basel in Switzerland. In 1994, he discovered an eye gene that was shared by fruit flies, mice, squid and humans. That caused great excitement because it suggested that the eye first developed in a common ancestor.

Gehring created bizarre flies that sprouted eyelike structures in places such as wings, legs and antennae when he fiddled with the gene – which is known as 'eyeless' because it was discovered when a mutated version of the gene stripped the insects of their eyes. He then found that an equivalent gene in mice, called Pax-6, could also cause fruit flies to grow fly eyes. The same went for squid. It does not seem a great leap to say that Pax-6 was probably used to enable Nifflers to hunt down the bright, the shiny and the sparkling.

Another gene involved in eye development pays homage to Argus Filch, the Hogwarts caretaker. Filch gets his name from Argus who, according to Grecian myth, had 100 eyes. While working at the University of California, Berkeley, Matthew Freeman named a gene that he had found after Argus (he actually used the Greek form, Argos) when he discovered that the gene was responsible for a protein that blocks the pathway of a growth hormone receptor, called the EGF receptor, which is involved in many stages of fly development, including that of light-sensitive cells in the eye. This discovery is of more than academic interest because, when overactive, the EGF receptor causes tumours in humans, so Argos points towards one way of shutting this cancer-promoting receptor down. Like so much of science, Freeman's research seems at first sight to be, at best, esoteric and, at worst, as pointless as a Muggle trying to find a Moke. (For those who have not read Scamander, a Moke can shrink at will so it is hard to spot.) Yet the quest to reveal the secret of Argus Filch and the Nifflers could uncover new ways to treat cancer. This is another reason that science is magical. No one can ever predict what will emerge from human curiosity.

Hippogriffs, geeps and chimeras

Myth has seen many creatures straddle the barrier between different species. The ancients created the legend of the mermaid, a fishy human; the satyr, who was half man/half goat; and the chimera, a fabulous beast made up of a lion's head, a goat's body and a dragon's tail. Another hybrid of known creatures is the winged horse of Greek mythology, the Hippogriff, from the Greek *hippos* (horse) and Italian *grifo* (griffin) which was supposedly a symbol of love. Well, at least to the Italian poet Lodovico Ariosto.

Modern science is beginning to bring to life such chimeras, which can be created by mixing embryos – embryonic stem cells – from different animals. The 'geep', for example, produced in Cambridge, England, in 1983, has goat-like horns and a partly sheep-like coat. The project sounds frivolous, but the head of the laboratory, Chris Polge, explained that the team was interested at the time in what enables a foetus to be tolerated by the mother during pregnancy, an immunological issue that sheds light on what can go wrong, leading to miscarriage.

While a sheep will not tolerate a goat embryo and a goat will not tolerate a sheep embryo, both animals will tolerate a geep embryo. This explains another rationale for this research: it is possible to create a chimera so that a common species could give birth to a foetus from an endangered species. 'You can make the chimera so that the placenta forms from one set of tissues [from the surrogate mother], and the foetal tissue from another [the endangered species],' says Polge.

There are many other chimeras. In 1996 the black-and-white 'Mixy' mouse was developed at Cambridge using new stem-cell technology that mixes cells. The following year a chicken that thinks it is a quail was created by moving brain tissue from one species to the other. The confused bird was created by Kevin Long and Evan Balaban at the Neurosciences Institute in San Diego, California, by transplanting the anterior midbrain from quail to chick. The resulting 'chimeric' chick responded preferentially to a quail mother and wobbled its head up and down like her too. Though it postured like a chicken, it engaged in the distinctive three-note quail song. This work can shed a little light on one of developmental biology's nagging questions: Where in the brain are the cells controlling specific behaviours located?

There have even been human chimeras. Suzanne Ildstad of the University of Pittsburgh and her colleagues replaced 20 per cent of baboon bone marrow with human stem cells, so the baboons had various lineages of human blood cells flowing through them. A mixture of human and baboon blood cells was attempted in a radical new AIDS treatment. In December 1995 Jeff Getty, thirty-eight, underwent the baboon marrow transplant at San Francisco General Hospital in the hope that he would begin producing disease-fighting baboon immune system cells. The treatment did not work.

How Fluffy got a head in life

Hagrid bought this monstrous three-headed dog with mad rolling eyes from a 'Greek chappie' that he met in the pub. This is presumably a reference to the origins of Fluffy in Greek mythology: Cerberus, the three-headed canine that guards the infernal regions and which, as Orpheus discovered, could be tamed by playing a lyre. We have already discussed how to create a huge hound; fortunately, adding an extra head is easy, according to Jim Smith of the Wellcome Trust/Cancer Research UK Institute in Cambridge. Additional heads do sometimes sprout in nature spontaneously, presumably as a result of a mutation: recently, Spanish researchers became excited by the discovery of a two-headed ladder snake.

Smith points out that a Nobel Prize was awarded to the late Hans Spemann for planting new heads in his work on experimental embryology. Spemann became a master of microsurgical techniques and, in 1924, working on the relatively large eggs of amphibians, discovered together with Hilde Mangold something that they called the organiser. When Mangold grafted the organiser – tissue – from one frog embryo to another, the result was a frog with two heads.

Recently, scientists discovered some of the genes that are used by the organiser, of which one has been fittingly named Cerberus. What is interesting about Cerberus is that it does not tell cells in a body specifically to start building a head, as one might expect, says Smith. Instead, it *inhibits* the signals that instruct cells to make a trunk. When that happens, a head emerges. 'It is bizarre,' says Smith. 'The head is what forms if you suppress the signals that build the rest of the body.'

To turn on Cerberus to form a second head in a frog, for example, would require an injection of Cerberus genetic material into a frog embryo. However, to achieve the same result in a dog or other mammal is more tricky. Smith explains that you would have to find a genetic switch, called a promoter, that is turned on only in the part of the trunk where you wanted to grow the extra head. This promoter could then be activated to in turn switch on Cerberus to create another head. The package of promoter and Cerberus could be introduced into a canine host by using a virus or by creating a genetically modified dog. However, building an animal with three heads is more difficult, not least due to space constraints. Nonetheless, Smith concludes that, ethical objections aside, it is feasible to create creatures such as Fluffy, double-ended newts, or Sisiutl, a double-headed snakelike being that, according to legend in the Pacific Northwest, strikes terror in human hearts, turning those who cannot face fear into stone.

Recently, scientists found a way to manipulate molecules to create chicks that have two beaks, marking the first time that facial features have been altered in a reproducible way without moving tissues around. The chicken experiments by Joy Richman and Sang-Hwy Lee, at the University of British Columbia, Vancouver, involved blocking a protein that stimulates bone growth and adding a chemical relative of vitamin A, called retinoic acid. This altered the patterns of bone and cartilage growth in developing chicks and changed the fate of one region of the embryonic face; cells destined to become the side of the beak were transformed and sent down a different developmental path to become the centre of a new beak. This suggests that concentrations of bone-growth proteins and retinoic acid probably determine facial-feature programming. Because nature is so parsimonious, the same signals are likely to be used during human development to specify the different parts of the face. Leaving aside the tricky issue of how to grow Voldemort's brain in Quirrell's head, it is at least possible to see how one man could end up with two faces.

Such experiments are not a frivolous use of scientific magic. The work is motivated by a need to understand the genes that control facial development and that, when faulty, cause deformities. Genetic mutations that affect this part of development are rarely lethal but often ruin a person's life. 'Our goal as scientists isn't to create monsters but to learn more about normal development,' says Richman.

Regrowing bones

Many bones are snapped in the Harry Potter books. Think of poor Arabella Figg, the mad old lady who broke her leg when she tripped over one of her cats. Neville broke his wrist during his first broomstick flying lesson. Harry, of course, fractured his right arm when a charmed Bludger crashed into his elbow. Professor Lockhart made matters worse by spiriting away the bones from Harry's arm with a dud spell. Fortunately, fixing broken bones is easy at Hogwarts, for Matron Madam Pomfrey simply reaches for her Skele-Gro.

One scientist who thinks he understands Lockhart's spell, and the restorative power of Skele-Gro, is Clinton Rubin, director of the Center for Biotechnology at the State University of New York, Stony Brook. Rubin has invested a great deal of time in reviewing the Harry Potter books with his young son Jasper and now considers himself an expert.

First, how did Harry lose his bones in the first place? The wave of the wand in Lockhart's spell could have activated 'bone-eating' cells called osteoclasts, explains Rubin. These cells are known to play a role in bone diseases such as osteoporosis and the loss of bone that takes place when people live in low-gravity conditions. This rapid bone loss (which admittedly takes longer than the wave of a wand) is one of the biggest hurdles facing astronauts on long-duration space missions.

Lockhart may have accelerated the process of bone loss with agents that stimulate the growth of osteoclasts. These have suitably spellbinding names, such as MCSF, RANK-L, or parathyroid hormone. Another way to erode bones is to deprive someone of sunlight, which in turn switches off the processes in the body that make Vitamin D, which is involved in calcium metabolism.

Fortunately, research that has shown how to lose the bones in an arm also suggests ways to regrow them. One method would be to activate the osteoblasts, or bone-forming cells. 'Again, in the spirit of Harry Potter, you could do that with low-level electrical fields, as used in orthopaedic clinics to stimulate broken bones to heal again,' says Rubin.

As for Skele-Gro, Rubin suggests that it takes advantage of proteins that have been shown by molecular biologists to be involved in the process of bone formation: 'A favourite candidate comes from the TGF-β superfamily, known as BMP, or Bone Morphogenetic Protein.' When BMP is introduced into the body it can stimulate the formation of cartilage and bone (and, when inhibited by the Cerberus gene that we encountered earlier, it contributes to head formation). Other factors, such as IGF-1, might do the same. And there are even some proteins with the power to make bones both wane and wax, offering a way to dissolve them and build them up. 'PTH, if given at high doses over time, would cause bone to resorb,' says Rubin, 'while PTH given in one shot, once per day, would stimulate bone to grow.' The result: one arm, reboned and very stiff.

Finally, there is a vibrant way to boost bones with none of the throat-burning side effects reported by witches and wizards who use Skele-Gro. Rubin has found that a shake-up strengthens bones: in experiments in which he vibrated the hind legs of sheep thirty times each second for twenty minutes each day, he produced a 30 per cent increase in the density of bone tissue, compared with controls. It is known that exercise boosts bone strength and these subtle vibrations may have the same effect because they mimic naturally occurring vibrations in the contractile elements of muscles at between 20 and 50 times per second (Hz) that take place thousands of times each minute for hours at a time. A similar approach, relying on almost imperceptible vibrations that are still ten times greater than those usually delivered by muscle, could offer a drug-free treatment for elderly osteoporosis patients. In that way, he hopes to improve the bone structure of patients who are too unfit to exercise and may, due to poor muscle tone, lack those bone-boosting muscle vibrations. Even in a person like Harry, who has a normal musculature, Rubin believes he can improve bone mass. If Madam Pomfrey ever runs out of Skele-Gro, Rubin believes he could offer her a slower alternative that would give her patients a therapeutic buzz.

Kappas

Kappas, creepy Japanese water dwellers that look like monkeys with webbed hands, could provide a glimpse of what Muggles would look like if a deadly process was turned off during our development. What is fascinating is that all of us, even Harry Potter, start out with webbing between our digits but then lose it due to a process called cell suicide, programmed cell death, or apoptosis (Greek for 'falling autumn leaves').

Without cell death, there can be no life. The discovery that every cell in our bodies stands ready to kill itself at a moment's notice marks one of the most profound recent advances in medical understanding. The idea was originally put forward in 1972, when Edinburgh University's Andrew Wyllie, Alastair Currie and John Kerr suggested that cells in the body have a suicide programme that they switch on when their continued existence would be harmful. For two decades, the scientific community sniffed: 'So what?' Now apoptosis is one of the biggest fields of investigation in biology.

Your body calls on cell suicides for the greater good. Every hour, around one billion cells do themselves in to maintain your shape and eliminate surplus or damaged cells. The process is orderly. Suicidal cells neatly break their chromosomes into tiny pieces and demolish all their internal structures. The cell membrane collapses to package these contents into tiny 'bin bags' without spilling toxic trash that could harm neighbours. The bags are collected by nearby cells for recycling.

Study of apoptosis has provided insights into development. For example, the male nematode worm kills 148 cells during the growth of its millimetre-long body. The field also sheds new light on cancer, which is linked to a failure of the suicide programme to weed out malignant cells. If cells die when they should not, the result may be hair loss, or even worse – AIDS or a degenerative illness such as Alzheimer's. Cell suicide is also used by the body to maintain overall cell numbers after healing and in its embryo stage to 'sculpt' tissues into a more mature form, for instance to remove webbing between foetal fingers and toes. Which brings us back to the kappas, which have found a way to suppress that particular suicide programme.

The real GM

There are many other examples in which GM – genetic magic – may have been used to create the odd creatures living in and around Hogwarts. Fawkes the phoenix, with his red and gold plumage, has healing tears that come to Harry's rescue, for example. Leaving aside the origins of the bird himself, evidence already exists that tears do indeed have healing ingredients, such as proteins that stave off bacteria. Indeed, Sylvia Lee-Huang of New York University and Hao-Chia Chen of the US National Institute of Child Health and Human Development found that lysozyme, a well-known protein abundant in tears, is a potent anti-HIV agent. Perhaps, by some well-targeted GM, Fawkes has been given tears with turbocharged therapeutic potential.

There is also a hint of genetic magic at work in Polyjuice

Potion, used to turn one person into another for a short while. A molecular biologist may well be struck by how the potion requires a sample of the person to be reproduced, suggesting that somehow, the potion exploits the genetic recipe – genome – of that person to reprogramme the cells of whoever quaffs the enchanted brew. Given how profound a process it would be to exchange the genetic code in a body, it is surprising that Harry only reported a burning feeling, a melting sensation and his skin bubbling like boiling chocolate. Because the cells in his body were being reprogrammed, one would have expected it to feel much worse than this. Presumably, however, his brain remained untouched so he did not think and act like Goyle, the DNA donor. Alas for Hermione, however, who started out with the DNA of a cat – the Polyjuice Potion left her feeling feline.

Molecular biology can do more than shed light on these transformations and the peculiar creatures in the magical bestiary. It can also explain one of the greatest culinary mysteries of all: how does Bertie Bott manage to create any and every tongue sensation, from the great flavour of butterbeer to the sickly tang of congealed Bubotuber pus and the stomach-churning taste of rancid eels?

CHAPTER 7

Bertie Bott's Every Flavour Beans

How the sense of beauty in its simplest form – that is, the reception of a peculiar kind of pleasure from certain colours, forms, and sounds – was first developed in the mind of man and of the lower animals, is a very obscure subject. The same sort of difficulty is presented, if we enquire how it is that certain flavours and odours give pleasure, and others displeasure.

Charles Darwin, *On the Origin of Species*

When you pop one of Bertie Bott's beans into your mouth you take a gastronomic gamble. As the advertisement at the Quidditch World Cup warns, there is a risk with every mouthful. Each bean tastes different, and you may find yourself in gourmet heaven, hell or indeed anywhere between. These beans can have the sharp flavour of sherbet lemons, which Dumbledore adores. But they can also taste of curry, pepper, tripe, grass and even, horror of horrors, brussels sprouts.

Imagine that you have taken your first chew. Yuck, the stomach-churning tang of vomit mixed with the whiff of a Dungbomb. Hmmm, the bland flavour of porridge. Mmmm . . . the sensational butterscotch flavour of warm butterbeer, as if it had just been served by Madam Rosmerta in the Three Broomsticks. But in an instant it becomes clear that there is more to taste than what is happening in your mouth. It is as the bean's aroma wafts up into your nose that many important aspects of flavour fully blossom (as anyone with a streaming cold can't tell you).

After decades of study we now know that flavour is more than what happens on the taste buds (gustation) alone. Smell (olfaction) and 'mouth feel' of the beans also play a role – and even the colour of food can influence our perception of how it tastes. The development of Bott's multiflavour beans comes as no surprise to scientists in the field because of the amazing strides made in recent years in understanding what goes on in our mouth, nose and brain when we chew food. It is indeed now possible, in theory at least, to rustle up any flavour in the laboratory.

We used to describe taste in terms of four qualities: saltiness, sourness, sweetness and bitterness. Scientists now strongly suspect that there is a fifth dimension to taste. Called *umami*, it is the meaty flavour typical of glutamate, one of the twenty amino acids that make up the proteins in meat, fish and legumes.

The first contribution to our perception of flavour comes from the taste buds on the tongue and soft palate. The majority are located within the papillae, the tiny projections that give the tongue its velvety appearance. (However, the majority of papillae actually lack taste buds and serve to provide tactile sensation – the feel of food in your mouth as you chew it or roll it around.) Although it is often claimed that different parts of the tongue respond to different tastes, this is a misinterpretation of nineteenth-century research; there may be slight differences in sensitivity, but you can get all types of taste from any location with taste buds, which is reassuring for those of us who like to cram their mouths with beans.

Each taste bud consists of up to 100 taste cells. At the top of each one is the taste pore, through which fingerlike projections called microvilli poke. When Harry chews a bean, it releases chemicals (tastants) that can enter the taste pores. There, they interact with molecules on the microvilli called receptors. When the right chemical activates a taste receptor, it launches a cascade of molecular events in the taste cells that culminates in the brain's perception of a flavour.

What happens at the molecular level, for example, when Harry chews a bitter bean, such as a Brussels sprout variety? The start of this process has been revealed by Charles Zuker of the University of California at San Diego, Nicholas Ryba of the US National Institute of Dental and Craniofacial Research, and Linda Buck of Harvard Medical School. They have discovered where in the body bitter tastes first register – the actual receptor proteins in taste buds that bind to bitter tastants. These so-called T2R/TRB receptors are part of a family of related receptors that is thought to have about twenty-five members. Experiments show that a receptor that reacts with one form of bitter tastant essentially ignores other bitter compounds. This helps to explain why there are up to twenty-five different bitter receptors, for no single receptor could possibly recognise so many variations. But why should we have so much molecular machinery to pick up a taste we don't like? The reason for this bitter-receptor overkill is survival itself. Virtually every naturally occurring toxin and poison tastes bitter. The perception of bitterness 'clearly evolved with the sole purpose of warning you', says Zuker.

Despite there being twenty-five different bitter receptors on our tongues, we do not distinguish among twenty-five types of bitterness or twenty-five variants of bitter Bott bean. The reason for this is that all twenty-five bitter receptors send their signal through one central protein called gustducin, which was discovered by Robert Margolskee of the Mount Sinai School of Medicine. Found inside taste cells, gustducin acts as a molecular switch that turns on the bitter-sensing taste cell when any of the twenty-five bitter receptors encounter broccoli, castor oil or any other kind of bitter substance. Genetically altered mice that lack gustducin contentedly devour the most awful bitter stuff as readily as they would eat cheese or chocolate.

Experiments with people show we are unable to distinguish one bitter substance from another. Our inability to tell underlines the survival benefit of this taste: it is important that we recognise *anything* bitter, thanks to our gustducin, but not to get hung up on distinctions among making different potentially hazardous compounds.

Progress is being made to find other receptors for other tastes. For those who like sweets, a number of groups – including Linda Buck's team and Robert Margolskee and colleagues and scientists at the flavour and fragrance company Senomyx – have found the potential sweet receptor gene product T1R3 (for 'taste receptor family 1, member 3'). Mice with a sweet tooth have one kind of T1R3, while mice that turn up their noses at sweet flavours have a different variety. Work by Danielle Reed and her colleagues at the Monell Chemical Senses Center in Philadelphia has found that people with a special form of T1R3 crave all sorts of sweets. Work by Zucker and colleagues suggests that T1R3 does not work alone, but with a second gene called T1R2. In a similar way, the umami receptor seems to be a combination of two genes, T1R1 and T1R3.

The flavour of Bertie Bott's beans also depends to a great extent on the interaction of hundreds of different molecules released by the beans into the air. These odorants dock with another group of receptor cells in the nose. The human olfactory system is capable of distinguishing between at least 10,000 different smells.

Once an odour has negotiated a hairpin-like turn at the top of the nasal cavity, it encounters the olfactory epithelium, a patch of cells containing about five million olfactory neurons that have specialised receptors that respond to smells. When the receptor proteins connect to molecules that have wafted into the nose, a cascade of chemical reactions is set in process that ends in the brain, where we register a smell.

But how, actually, do we smell? One way to detect the full gamut of aromas, from pumpkin tarts to cinnamon bogeys and curried earwax, would be to have a receptor designed to respond to each and every one. This would be very expensive on receptors given that there are 10,000 discernible smells. Since each protein is described by a gene, this would mean that around one-third of humankind's hereditary code, comprising some 30,000 genes, would have to be dedicated to making smell receptors. This seems unwieldy.

Alternatively, there could be just a few types of receptor. In this case, each would respond to many smells, reacting to each aroma in a slightly different manner. This set-up is more efficient and has a well-known precedent. Our eyes detect colour in a similar way: they have only three types of colour receptors. By comparing the responses of all three, the brain paints relatively few nerve signals into all the hues of the spectrum. It turns out, from work by Linda Buck and her colleagues, that each neuron in the olfactory epithelium makes only one of the 1,000 or so types of olfactory receptors on its surface.

There is a complex relationship between the smelly molecules released by a bean and the triggering of receptors. A single receptor can recognise several odorants; a single odorant is typically recognised by multiple receptors; and different odorants are recognised by different combinations of receptors. In other words, different combinations explain how hundreds of receptors can describe many thousands of odours.

Even slight changes in the chemical structure of odour chemicals can activate different combinations of receptors. Thus, octanol smells like oranges, but the close chemical relative octanoic acid smells like sweat. Similarly, large amounts of a chemical bind to a wider variety of receptors than do small amounts of the same chemical. This would explain why a large whiff of the chemical indole smells putrid, while a trace of the same chemical smells flowery.

Once an odour molecule has docked with a receptor on the olfactory epithelium, the signal travels to the olfactory bulb. This structure, located in the front of the brain, is the clearing house for the sense of smell, according to studies by Linda Buck and Bettina Malnic at Harvard Medical School, and colleagues in Japan.

Signals from neurons with the same odour receptors converge on tiny basketlike structures called glomeruli in the olfactory bulb, creating a map of odour receptor inputs. This map is nearly identical in different individuals. In this way, everyone can agree on the smell of peppermint, for example. 'It provides a potential explanation as to why the odour of, say, a skunk smells bad to all people, and roses smell sweet,' says Buck. From the olfactory bulb, odour signals are relayed to the higher cortex, the part of the brain that handles conscious thought. 'It appears that information from different receptors is being combined in the cortex, while that's not true in the nose or the olfactory bulb,' she relates.

Importantly, odour signals are also sent to the brain's limbic system, which generates emotional feelings, such as the 'yuck' that results from chewing a vomit-flavoured bean, or the 'ahhh' when one savours the taste of a childhood favorite. 'This is probably why an odour can evoke powerful emotional responses as well as convey factual information,' Buck explains, adding that there is probably a link with the amygdala – the structure involved in processing information associated with intense emotion, such as the smell of burning or of perfume.

With this reasonably detailed picture of how our sense of taste actually works, scientists are now in a position to find ways to mimic all kinds of flavours and even create some that are entirely new and strange. One company, Senomyx in La Jolla, California, is using the knowledge of receptors to find molecules that 'enhance the human sensory experience'. With an understanding of the sweet receptor and the way sweet flavours bind to it, scientists can design a replacement for sugar – a molecule that would fit the receptor perfectly and be a million times more potent than sugar, yet have the same sweetness as a natural sugar sweetener. Another company, Linguagen in New Jersey, is targeting the protein gustducin to block every kind of bitter taste. Their bitter blockers – such as a molecule called adenosine monosphosphate – should make coffee taste smoother and help the medicine go down more easily. These companies are also working on ways to tackle bad smells, boost the salty taste of low-sodium snacks and so on. To do this, they are combining the improved understanding of the biology of taste and smell with many of the same technologies that pharmaceutical companies now use to find new medicines.

However, their work has also shown that Bertie Bott's beans do not have the same impact they would have had in ancient times. The first analysis of the human genetic code suggested that there are around 1,000 olfactory receptor genes. Recently, Sergey Zozulya and colleagues at Senomyx found that only about 350 of them are functional (a number that probably varies from person to person), with the remainder being 'dead', so called pseudogenes, the remnants of once-working genes that have been wrecked by mutations. One study suggests that the number of these dud odour genes has increased in number from monkey to man. As a result, our repertoire of odour receptors is now about one-third the size of that of rodents or dogs, says Zozulya. We have an abridged – 'thinned out' – version of the same repertoire used by mice and presumably other mammals.

The most plausible explanation for this attrition is that there was a decreasing biological need for a well-developed olfactory apparatus in our ancestors. Animals continue to rely on their keen sense of smell to find food, detect the presence and physiological state of their kin and other species, avoid predators and other dangers, and so on. For humans, most of these functions are no longer very important. Given that the genetic codes of creatures are forever being restructured, mutated and shaped, if a gene is not needed for survival or a significant selective advantage, it will eventually be lost. 'It is a "use it or lose it" situation,' says Zozulya.

The best and worst flavours

From a Darwinian perspective, our tastes were honed to ensure survival. In the womb, embryos increase their intake of amniotic fluid when a sweetener is injected, revealing an innate love of sweetness that prepares babies to enjoy breast milk and enables them to tell when fruit is ripe enough to eat.

Equally, babies hate bitter tastes because many naturally bitter products are harmful: for example, alkaloids such as strychnine are poisonous, and the ability to detect them by taste evolved as an adaptation for survival. Hence the one-taste-fits-all response to these toxins, discussed above, and the well-known childish aversion to bitter foods such as Brussels sprouts.

There are certain flavours that everyone likes (chocolate comes to mind) and everyone hates (vomit), but the candidates for the very best and the very worst tastes depend very much on the experience of the individual. We can of course train ourselves to like bitter tastes and sometimes do so for good reason. The bitter principles in 'acquired tastes' such as aperitifs, or the quinine in a gin and tonic for example, can help stimulate the production of saliva before a meal.

For her research, Dana Small of Northwestern University has had to give a lot of thought to the issue of a universally liked and disliked tastes – Bott's best and most bilious beans. While studying the brain regions underlying pleasant and aversive taste with a PET brain scanner, she had to find the opposite ends of the flavour spectrum. 'This was no easy task,' she says. Although we are born with a sweet tooth, 'It is near impossible to find anyone who will rate sugar water as extremely pleasant, although it is easy to find people who will rate bitter water as extremely unpleasant. This fact led me to my current thinking that the main role of pure taste, which has been described as the final gatekeeper to the internal environment, is to reject poisons.'

In the end, she selected chocolate as her standard pleasant taste, since it is the most craved food of all, at least among Western cultures. There is more to chocolate than taste alone. Other sensual effects contribute, such as its wonderful mouth feel and the seductive sensation of coolness as it melts. And when chocolate lovers eat chocolate they activate many brain regions also activated by addictive drugs.

But context is also important to how we relate to information about taste and smell, as the French novelist Marcel Proust demonstrated with lime tea and madeleines. The smell of burning wood is comforting if you are warming yourself in front of a blazing fire on a cold winter's night, but frightening if you are sitting in a darkened cinema.

This phenomenon was revealed by turning a brain scanner on women as they sniffed a variety of disgusting smells – rotten eggs, garlic breath or sewer gas – in experiments conducted by David Zald and José Pardo at the Veterans Affairs Medical Center in Minneapolis and at the University of Minnesota. The scientists were particularly interested in the effects of these odours on the almond-shaped amygdalae. Each half of the brain has one amygdala, and together they are a key part of the brain's machinery for creating emotions.

When the volunteers were exposed to the worst stench, the team could see the amygdalae being used, as if they were telling the rest of the brain, 'Hey, you really hate this stuff.' Pleasant smells, which included fruits, flowers and spices, evoked only a weak response in the amygdalae, by contrast. One woman provided a good example of the influence of context. She found that one supposedly bad odour really wasn't that terrible, and the scanner showed her amygdalae agreed. The reason: she had spent a vacation in Alaska near an oil refinery and the smell 'reminded her of that wonderful summer she had', Pardo explains. It is easy to see why everything from the fragrance used by a partner or that indescribable smell of a newborn baby could be candidates for the nicest smell of all.

Familiarity is also important when it comes to weighing up how much we like a taste or smell. Dana Small found that one Asian man who claimed to love chocolate found it impossible to cope with more than a few bites. He had only started to eat chocolate a few months beforehand and lacked enough experience to have become a chocaholic.

Chocolate is certainly liked by Noam Sobel of the University of California, Berkeley, but he most craves the smell and taste of 'my mom's coffee-peach-crumb cake', for obvious reasons. As for the worst smell/taste of all, he says that faeces is a strong contender. Once again, however, context is all important. Most people are not that disgusted by the smell of their own faeces, or by that of people close to them, said Sobel, who was once (almost) moved to investigate this fascinating observation. 'The hedonics of both smell and taste are strongly influenced by experience. The ripe or rotten smell of ageing cheese may be divine or sickening depending on whether you come from Paris, France, or Paris, Texas, respectively.' (The organic acids found in strong cheeses are also found in vomit and foot sweat.)

Being unfamiliar with the foul stench of decay caused by a Bundimun, a patch of greenish fungus with eyes, Sobel instead put forward burned-hair flavour as another candidate for Bott's worst bean. But: 'I know of a cross-cultural study that showed that while Europeans greatly disliked the smell of burned-hair, some African people associated the smell with traditional medicines and did not find the smell to be nearly as objectionable as did Europeans,' comments Jeannine Delwiche of Ohio State University who nominates as the worst flavour alum, an astringent substance consisting of aluminium, potassium and sulphate, which tastes 'pretty awful'.

We all recoil when exposed to the chemical constituents of horrible smells: hydrogen sulphide (bad eggs), methyl mercaptan and skatole (faeces), cadaverine (corpses), putrescine (decaying meat), isovaleric acid (sweaty feet). Vomit is also a very strong contender for the nastiest bean because its smell has been honed by evolution to have maximum impact. The merest whiff of vomit can make others retch and heave as a protective response. Vomiting is designed to make us expel toxic substances that have been eaten. Because people lived in small communities for most of human history, it is likely that if a toxin affected one person in a group, others would be struck down too. If someone started being sick nearby, it was a good thing that his tribemates also emptied their stomachs, for they had probably shared the same meal.

Other insights into the most horrible bean of all might come from a project by the United States Army to develop the mother of all stink bombs. The US military dreams of a stench that is so repulsive that it can quickly disperse rioters and empty streets. The effort to create a mixture of malodorous molecules that does not kill or maim but can halt disorderly mobs is part of the Pentagon's Nonlethal Weapons Program.

This is not the first time that the US military has become obsessed with odour. During the Second World War it conjured up 'Who Me?', a stench meant to help the French resistance humiliate German officers by making them smell foul. But it was so volatile that it was impossible to 'bomb' the target without contaminating everything nearby.

Pam Dalton, of the Monell Chemical Senses Center, has led a team that has created and tested a range of nasty smells on volunteers in a quest for what she calls a 'universal malodour'.

She found that two odours transcended culture. One was standard bathroom malodour, a faecal stink concocted to test cleaning products. The other was an updated version of 'Who Me?', a bouquet of sulphur molecules reminiscent of spoiled food and rotting carcasses. A combination of the two should affect everyone, she believes. There is probably a Bott's bean lurking out there that contains this gastronomic nightmare. Beware.

PART II

The history of thought should warn us against concluding that because the scientific theory of the world is the best that has yet been formulated, it is necessarily complete and final . . . In the last analysis magic, religion and science are nothing but theories of thought; and as science has supplanted its predecessors, so it may hereafter be itself superseded by some more perfect hypothesis.

Sir James Frazer, *The Golden Bough*

Foreword

No history of magic would be complete without a reference to the tragic state of the Muggle mind. The many shortcomings of this soft, convoluted mass of nervous tissue can explain much about witchcraft and magic through the ages and, in turn, Muggle accounts of monsters, sorcery and enchantments can tell us much about the flawed mechanism at work between every pair of human ears.

Muggles think they have an extraordinary encephalon. They endlessly remind us of how the human brain is the most complex thing they know of, but not of its manifest shortcomings. They are fond of telling us that they believe the evidence before their eyes, but fail to realise how that evidence is mangled by their muddled minds. They trust their memories. Yet their power of recall is contaminated by imagination and suggestion. Alas, dear reader, when it comes to mental ability, Muggles are quite a few twigs short of a broomstick.

There are many and various limitations of the Muggle mind, from delusions and hallucinations to manifest illusions. The poor things can become addled, muddled and befuddled, not to say bemused, confused and perplexed. Even a tad confounded. Inspired by fossils, or by the misfortune of others, they can conjure out of thin air a menagerie of creatures – dragons, werewolves, pixies and giants. Deprived of their full faculties, they begin to see witches, warlocks, demons and ghosts. Even by waking from a stupor, they can become convinced that an old hag is squatting on them. Moreover, the human memory is so flawed, and imagination so fertile, it is a wonder why wizards bother with Memory Charms at all when Muggles catch a glimpse of anything real from the wizarding world.

More evidence of their mental shortcomings can be found in the quite dreadful treatment that Muggles have meted out to wizards and witches in the past, from torture and hanging to stonings and burnings. This is all the more striking because not one victim that I know of was any good at sorcery. What makes Muggle minds believe in the peculiar things that they do that drive them to carry out these awful acts?

There's more. Among the extraordinary convictions that infect the Muggle mind is a belief that they are on the verge of a theory that can explain everything. But no Muggle is ever going to explain why Harry Potter likes butterbeer, Hagrid's love of lethal creatures, and a trillion and one other things. There is good reason to be confident that the view of Muggle science as the cold-eyed, mathematical enemy of wonderment is wrong. Science will never be able to erase all the magic from life.

CHAPTER 8

Stars, mystic chickens and superstitious pigeons

Superstition is the religion of feeble minds.

Edmund Burke

Hundreds of years ago, there was not a clear distinction between science and what we would today consider paranormal pursuits. At the start of the seventeenth century, influential figures such as Tycho Brahe, Galileo Galilei and Johannes Kepler – all celebrated for their roles in the development of modern physics – held astrology in high esteem. The great Irish pioneer of chemistry, Robert Boyle, believed in angels and the ability of the philosopher's stone to make gold. Sir Isaac Newton, perhaps the most influential figure in the history of science, had a lifelong interest in alchemy and religion.

Although there is no explicit science in the Harry Potter books, there is good evidence that the boundary between science and magic is blurred at Hogwarts, as was once the case in the Muggle world. Our green-eyed hero has to observe the night skies every Wednesday, and his astronomy teacher Professor Sinistra teaches him about the movement of the planets and the names of different stars. So far, so rational. But Professor Sybill Trelawney also reveals to him the supposed secrets of astrology along with palmistry, crystal orb gazing and poring over bird entrails in her divination classes. Harry studies herbology with Professor Sprout, and no doubt a little botany and gardening, too. His potions lessons with Professor Snape probably cover what Muggles would call chemistry and pharmacology as well as the usual hocus-pocus. As we have already seen, anything from genetic engineering to quantum physics could have come up during Harry's more unusual pursuits, such as transfiguration, defence against dark forces, fantastic beasts, charms, curses and spells.

Although a magical feat suggests something supernatural is going on, extending beyond the laws of nature, there are links between the practice of magic and that of science. Both magic and experimental science are spurred by curiosity about the world – how to gain knowledge of it and how to manipulate it. Both are performed by humans to deal with uncertainty and build confidence in a hostile environment.

For the ancient mind, magic may have provided a much needed sense of predictability. One answer to the ever-present challenge of survival was to carry out a ritual, or utter a spell or a prayer. By declaring what you wanted and acting out your deepest desire, you might actually make it happen. Magical ideas include the notion that thought can affect matter and that the trained imagination can alter the physical world, that all aspects of the universe are interdependent and that the experienced observer will be able to discover connections and correspondences between cosmic things and everyday events.

Science also looks for connections in nature so as to understand and manipulate the natural world. However, science goes one crucial step further than magic, putting these correspondences to the test in experiments to refine and evolve theory. Although it took centuries for science to gain ascendancy over magic, it has far from purged modern culture of fascination with the topic, as the current interest in Harry Potter suggests. There remain, in fact, points of contact between the scientific and magical domains. One, the need to believe in something, no matter how strange, is dealt with in Chapter 14. Here I discuss the faulty foundations that these beliefs are built upon.

The origins of superstition

Superstition, science and magic all thrive on the human quest to make sense of our surroundings. To do so, the mind was honed by evolution to seek patterns, such as those that concerned food and survival: the place where tasty and nutritious roots grew each year; the time that plants flower and fruit; and evidence that a fearful predator was stalking the neighbourhood. Many features and events in the world are organised in a non-random way and it is crucial to us, and to animals generally, to detect nature's regularities.

We are born pattern seekers. Babies can learn to recognise their mother tongue in the womb, can distinguish a foreign language within days of being born and are intrigued by magic tricks at 2.5 months. This overwhelming urge to seek patterns, such as the link between cause and effect, is the lesser of two evils. If we had never evolved to seek patterns, we would not be here to discuss the subject. The difficulty lies in getting the balance right between, on the one hand, failing to spot a pattern that could save one's life, and, on the other hand, being distracted from the important business of survival by looking for a supposed pattern where there really is none.

The facility is so important that the modern brain is an over-zealous pattern seeker in its attempts to link cause and effect: it is likely to construct coincidences, even if two events are not linked in any way. This is the source of much Muggle superstition and magic.

We are not alone in this propensity. Cats link events that occur close to each other, rather than discern true cause and effect, according to classic experiments performed half a century ago. At the University of Washington, Seattle, Edwin Guthrie and George Horton rigged a 'puzzle box' so a cat could engineer its own release by pushing a post. After observing 800 escapes by fifty or so cats, they discovered that the animals linked the last thing they did with escape, rather than a push of the release mechanism. Guthrie remarked: 'When escape from the puzzle box followed almost any behaviour, colliding with the release, pawing it, backing into it, jumping to the top of the box and falling on the release, lying down and inadvertently rolling to contact with the release, heavy odds could be placed that the same movement would be repeated soon after the cat was returned to the box.' The breakouts were described in the 1946 classic *Cats in a Puzzle Box*.

To be fair to the cats, they at least established a connection between escape and the release mechanism, even if they did not figure out that all they had to do was push it. An experiment can go further than this and be designed to erase all link between cause and effect, at which point coincidences can give rise to superstition. This was revealed by behavioural psychologist B. F. Skinner of Indiana University in the 1940s. Of all the articles that he wrote, one of the most influential appeared in the *Journal of Experimental Psychology* and was entitled '"Superstition" in the pigeon'.

In his classic experiment, Skinner first ensured that his pigeons were hungry and then placed the birds into a cage where food was delivered every fifteen seconds by a hopper. All the birds had to do was wait for the food. Yet the birds developed distinctive rituals, presumably because they connected them with obtaining another snack. 'One bird was conditioned to turn counterclockwise about the cage, making two or three turns between reinforcements. Another repeatedly thrust its head into one of the upper corners of the cage. A third developed a 'tossing' response, as if placing its head beneath an invisible bar and lifting it repeatedly. Two birds developed a pendulum motion of the head and body . . . The bird happens to be executing some response as the hopper appears; as a result it tends to repeat this response.'

The pigeons were alert to the possibility of links between events in their world but made a false causal connection: they were superstitious pigeons. We have much in common with them when it comes to being duped by false patterns and coincidences. Skinner himself remarked on the analogies: 'Rituals for changing one's luck at cards are good examples. A few accidental connections between a ritual and favourable consequences suffice to set up and maintain the behaviour in spite of many unreinforced instances. The bowler who has released a ball down the alley but continues to behave as if he were controlling it by twisting and turning his arm and shoulder is another case in point.'

Some have pointed out that these kinds of experiments may say more about the natural behaviour of the creature under study – notably, the tendency of a cat to rub its head against something, whether a convenient post that turns out to be an escape mechanism or an experimenter's leg. However, this kind of experiment has even created superstitious behaviour in children. In 1987, instead of a food hopper, Gregory Wagner and Edward Morris at the University of Kansas used a mechanical clown called Bobo, who dispensed marbles from his mouth. Each child who took part in the experiment was told that, if he collected enough marbles, he would be given a toy. As in Skinner's experiment, Bobo regurgitated marbles at regular intervals, regardless of what the child did. And just as the pigeons had done, the children developed superstitious attempts at control, whether grimaces, wriggling or touching Bobo's nose. The same year Koichi Ono of Komazawa University in Japan used an analogous approach to create superstitious behaviour in university students, rewarding them with points rather than marbles. One hapless participant thought that the secret of success was to jump and touch the ceiling ('She stopped about twenty-five minutes into the session, perhaps because of fatigue').

This research tells us that few people are content to accept that blind chance plays a large part in their lives. They seek reasons, correlations and logical connections – even when these do not really exist. The human mind, far from being infinitely malleable, tends to impose certain built-in expectations on experience. And whenever there is uncertainty and anxiety, there is more pressure to seek patterns, and we are consequently more likely to be duped by what seems like magic.

The magic of chance

Many people, not just the most gullible, would be impressed by a television programme during which viewers' watches stop when a psychic tells them to. However, with ten million viewers, someone's watch is bound to stop somewhere due to any number of humdrum causes, from a failed battery to a slack mainspring. Even if the probability of its occurring were one in a million, it would give the producers of the show ten people who, no doubt stunned by the psychic's power, would be moved to announce the miracle. By the superficial standards of television, that counts as proof of the psychic's ability. The wizard of statistics, Sir Ronald Fisher, once remarked: 'The "one chance in a million" will undoubtedly occur, with no less and no more than its appropriate frequency, however surprised we may be that it should occur to *us*.'

Many of us have had that unsettling experience where a dream or a brief thought about a close friend is followed by a knocking on the door, or a call to say hello or, perish the thought, by the news that he had died in an accident. One can explain this with the laws of statistics rather than occult forces, psychic connections or even those ever-present vibrating resonances of cosmic energy that criss-cross the ether. Given how often we think of other people, it would be more surprising if no one experienced a startling coincidence at some point during a lifetime. We tend to forget all those boring instances when we thought of someone and nothing happened to him. But, eventually, Uncle Freddy is bound to knock on the door seconds after you wondered how he was doing, and a friend or relative is bound to drop dead on the very day you thought about her.

Such coincidence becomes exaggerated in our minds because humans are masters of selective memory. We focus on coincidences, just as the lovelorn focus on the good times of a failed relationship rather than the misery. This tendency makes us receptive to magic.

Stone Age sorcery

The word 'superstition' is derived from the Latin *superstitio*, 'to stand over a thing in awe or amazement', which points to another reason that we are so superstitious: we are easily astonished. Richard Dawkins of Oxford University believes that our Stone Age brains are unduly amazed – and thus unduly superstitious – because they were honed for the requirements of our ancestors, who lived mostly boring, uneventful lives in small populations.

Anthropology and fossil evidence suggest that, for much of the past few hundreds of thousands of years, our ancestors probably lived in small roving groups or in hamlets. Thus coincidence had to involve a few dozen people at most. 'Our brains became calibrated to detect patterns, and gasp with astonishment at a level of coincidence which would actually be quite modest if our catchment area of friends and acquaintances had been large,' says Dawkins.

Today, of course, thanks to television, the internet and other forms of mass communication, our catchment area has become enormous, and that hair-trigger calibration within our brain that makes us gasp at a coincidence is engaged much more than when we were hunter–gatherers. 'Our brains are calibrated by ancestral natural selection to expect a much more modest level of coincidence, calibrated under small village conditions,' says Dawkins. 'So we are impressed by coincidences because of a miscalibrated gasp threshold.'

Another factor that may reduce the threshold at which patterns and coincidences have the power to impress is the pace of life. Modern life is frenetic and we have more experiences, travel further and do more things per hour than were possible centuries ago, boosting the number of opportunities for happenstance and further confounding the ability of our brains to weigh the significance of a coincidence.

The mystic chicken

With the urge to link cause and effect in everyday life remaining as strong as ever, the result is superstition – lots of it, mad, glorious, daft and everywhere. Stuart Vyse, an American psychologist, cites the example of Wade Boggs, the former New York Yankees third baseman, who became convinced that he performed better after eating chicken, particularly his own recipe for lemon chicken. He dined on the fowl before games for twenty years. He also jogged to the dugout in four strides, never trod on the foul line when running on to the field and, when he stepped into the batting box, drew the Hebrew symbol *chai* in the dirt with his bat.

The same thinking is adopted by the woman who wears toad-green shoes the night she wins at bingo and becomes convinced that her footwear works like a lucky charm. The same philosophy is adopted by the craps player who shouts out the desired number when releasing the dice, or the student who will only sit for an exam wielding his 'lucky pen'. We cannot easily shake off our basic survival instincts and still behave like Skinner's pigeons. Even if the woman with the green shoes does not win at bingo again, she will try to rationalise why the footwear failed to deliver. Perhaps she didn't tie the shoelaces right.

Many still believe that there is a link between walking under a ladder and bad luck, or a rabbit's foot and good luck, and many of us are still wearing charm bracelets and crossing fingers. Stuart Vyse once found that he and another psychologist were the only ones willing to sit in the thirteenth row of an aeroplane when every other seat was taken. 'Do people really think the thirteenth row will have a fate that is different from the other rows?' he wonders.

The building where I work, the tallest in Britain, does not have a thirteenth floor (it does, of course, but it is not labelled as such). Presumably it was designed either by someone who fears this number, a triskaidekaphobic, or with triskaidekaphobics in mind. The receptionist of one Boston hotel I stayed in told me that guests did not like to stay in room 666 (for good reason, I discovered. The room throbbed to the sound of ice machines and air-conditioning). Nearby, in one of the great powerhouses of American science, the Massachusetts Institute of Technology, generations of students have walked past the bronze plaque of inventor George Eastman and rubbed his nose for good luck. Easy to do and, you never know, it might just help.

Vyse says that many people regard superstitions as an insurance policy. They are a kind of 'Pascal's wager', referring to how the seventeenth-century French philosopher Blaise Pascal believed that it was rational to believe in God, just in case He exists. Even if there is only a slim chance that heaven and hell were real, one should lead a Christian life to protect against a risk of damnation. In the same way, many people reason that they have nothing to lose if they walk around a ladder, rather than underneath it.

Such superstitious acts are all the more remarkable because today many would insist that science has succeeded and magic failed. The reason is that scientific ideas are subject to scepticism (manifested by double-blind experiments, mathematics, placebos, reproducibility, peer review and other trappings of the scientific method). The result of an endless dialogue between theory and experiment is a body of objective knowledge and, most satisfying of all, complementary knowledge, so that what chemists know about the structure of the genetic material of inheritance (DNA) fits with what molecular biologists know about the units of inheritance (genes) and what zoologists see crawling, wriggling and flying around in the natural world; what computer models are revealing about evolution, and what fossil hunters find in dirt and deposits.

However, the gap between scientific understanding and the material benefits to be derived from it was long, obscuring the link between cause and effect. It would take centuries before it was clear to anyone that the 'scientific' physician could do more to cure illness – with a tablet of penicillin, say – than could the equivalent of Snape with his herbs and potions and muttered charms. It would also take centuries to disentangle astronomy from astrology and its supposed ability to judge the occult influence of the stars on human affairs.

Astrology

Wander up to the top of the North Tower of Hogwarts and you may be lucky enough to join Sybill Trelawney's divination class. Otherwise, try to get hold of a copy of *Broken Balls: When Fortunes Turn Foul*, or *Death Omens: What To Do When You Know the Worst Is Coming*. Among methods Trelawney uses to glimpse the future, from fire omens to palmistry, is astrology, where events can be forecast by observing and interpreting the stars, the sun, the moon and the planets.

Today, astrology is part of the entertainment industry. Woe betide anyone who calls Britain's chief stargazer, the Astronomer Royal, the Astrologer Royal. He will tell you, quietly but firmly, that astrology is bunkum. However, before the seventeenth century, there was not the sharp division between astronomy and astrology that we see today. In the dark days before scientific enlightenment, they were bedfellows. While astronomy focused on how the night sky changed from season to season, the movement of the sun, the moon and the handful of known planets, and the possibilities of predicting their positions, astrology interpreted their motions as a key to human characteristics and activities.

Our fascination with the night sky is in fact as old as human history. Many ancient beliefs in the power of the heavens emerged from the false causality illustrated by Skinner's pigeons. Take the Aztecs, who, concerned that the sun might not rise the next day, ensured the new dawn with a ritual evening sacrifice. Not once did they fail to get the result they wanted. Our ancestors held winter festivities of the kind seen at Hogwarts each Christmas to usher in the annual return of sunlight, warmth and fertility with rituals involving the evergreens which seemed to defy the cold winter months, and the yellow light of a living flame. In various pagan rituals a tree was decorated to encourage the 'tree spirits' to return to the forest so it would sprout again, which of course it did every spring.

But some very real and important associations emerged as ancient civilisations learned how to link heavenly patterns with survival. This interest in the heavens conferred an – albeit limited – ability to a wise man, holy man, shaman or whatever to foretell the future, guiding him through the seasons, revealing when to harvest and when to move herds. It also helped him to predict striking events, such as the flooding of rivers, as was the case with the Nile. In ancient times, anyone who could predict the darkening of the skies, when an eclipse took place, would command great power and respect (as Tintin once discovered in his adventures, and which was earlier exploited by H. Rider Haggard's *King Solomon's Mines*).

The sun, our local star, does indeed influence our lives. There can be no doubting how the gradual changes in the earth's orientation to the sun alter our climate, and there are seasonal changes in the incidence of disease and cycles of hormone levels in the body. In this very restricted sense, knowledge of the heavens does shed a faint light on our destiny.

Harry Potter's birthday can tell us a little about his future. Given that the five hundredth deathday of Nearly Headless Nick took place when Harry was twelve, and that N-H-N died in 1492, Harry was probably born in 1980. Clues in the books, notably a story in the *Daily Prophet* about the break-in at Gringotts Bank, led many to believe that Harry was born on 31 July, which also happens to be J.K. Rowling's birthday. (Not every clue agrees, however. The first book says Harry had his eleventh birthday on a Tuesday, when 31 July 1991 was a Wednesday.)

Harry is a Leo and is supposedly confident, brave and so on like the other one-twelfth of the global population who share this star sign. The reasons are unclear, except that this sign of the zodiac is also called the Lion. So far, so silly. However, recent research has given some fascinating insights into how his birth date could exert an influence for the rest of Harry's life, from the functioning of his immune system to his risk of heart disease to his intellectual and sporting achievement.

At the risk of putting Professor Trelawney out of business, let's see what may be in store for Harry. One weak 'astrological' effect arises in sports, where the development of the body matters. According to a study at the University of Amsterdam, the youngest children in any age grouping, such as Harry, are at a disadvantage because they are born late in a selection year.

This does not seem to have caused Harry any problems when it comes to Quidditch. Michael Holmes of Queen Margaret College in Edinburgh found evidence that May to September is the birth season for reactionaries. Another study showed that the summer-born are smarter, although the author recoiled from astrological explanations and speculated that babies born during warm months are less likely to be wrapped up, suffer illness or to be cooped up indoors. This would enable them to be more adventurous and enjoy a more stimulating environment, which has been shown by experiments on animals to be important for brain development. Now that does sound more like our green-eyed hero.

Beyond these weak connections between birth date and the future, the idea that astrology can say much more about life and personality is as preposterous as Professor Trelawney's gloomier pronouncements. Harry himself wisely came to the conclusion that she relied on lucky guesswork and a spooky manner. But it is unwise to write her off completely, however. Trelawney routinely predicts the death of a student; she is bound to be right one of these days.

Many millions of people believe that the position of the heavenly bodies at the time of one's birth helps determine one's personality and future destiny. No convincing mechanism through which the alleged planetary influences exert themselves has ever been specified by astrologers. Gravity cannot account for these natal influences, since the gravitational pull of the people nearby – doctor, husband and midwife – should be as significant. Nor is there any good evidence that astrology works: top astrologers (as determined by their peers) have failed repeatedly to associate personality profiles with astrological data at a rate higher than that of chance. Proponents usually quote the same old unconvincing studies and I, for one, have never seen anything supporting astrology in a science journal.

In this microchip age, when we can walk on that oh-so-influential moon and send probes to fearful Mars and beyond, many cling as strongly as ever to this pseudoscience. The idea that a distant object like Pluto can exert a measurable effect on our lives is absurd, let alone that an astrologer could predict it. True, the earth's relationship with our local star (the sun) is as important as ever, given its influence on the seasons. True, we can work out the movement of the planets with astonishing precision. True, quantum mechanics says that we can be affected by a single electron at the edge of the cosmos. But there are an awful lot of electrons out there. And don't forget that chaos theory also says that only with infinite information could we predict the future behaviour of even two billiard balls ricocheting around an idealised pool table, let alone something as complex as a human life. What is often not realised is that astrology is not only at odds with science but even with magic. The former assumes that humans are impotent, unable to control their destiny, while the latter assumes that humans are omnipotent, able to change it with the flick of a wand.

Weather and witchcraft

Wizards can tell when rainfall is coming by the mournful cry of the Augurey, also known as the Irish Phoenix. Unfortunately, its dour song is unknown in the Muggle world, where alternatives have been sought for millennia. Although the ancients could link the heavens with the passing of the seasons, a great deal of weather folklore is as old as legend and usually as unreliable. 'Take care not to sow in a north wind or to graft and inoculate when the wind is in the south,' said the Roman writer Pliny, in an adage opposite to one that circulated in the seventeenth century. However, not all ancient attempts to use the stars to look into the future can be written off, blurring the divide between science and superstition.

Studies of what should probably be called ethnoclimatology have shown that, when it came to using the heavens for long-range weather forecasting, the Incas in drought-prone regions of the Andes were star performers. A tradition handed down through generations of Incan potato farmers is that the brightness of the Pleiades, a cluster of stars that they worshipped, could predict rainfall several months later – from October to May – during the growing season.

For hundreds of years, local people linked a peak in the brightness of these stars – when there were clearer skies – to earlier and more abundant rains and larger potato harvests in March to May of the following year. The potato crop – locally the most important – is shallow-rooted and as a result is highly vulnerable to drought at planting time. Thus if poor rains were predicted by dim stars, the villagers in the mountains of Peru and Bolivia delayed planting for several weeks, starting the crop in months of higher rainfall.

The same star forecasting is still in use and was noticed by Mark Cane, a Columbia University climatologist, while he was on holiday in the Andes. 'A guide mentioned that a local farmer had forecast the upcoming harvest by going to a mountaintop in June.' Observations made by the farmers often began around 15 June and culminated on 24 June, the festival of San Juan. Cane speculated that their predictions worked because the visibility of the Pleiades was in some way related to El Niño, a major climate event that takes place every few years in the equatorial part of the Pacific ocean, when warm waters shift eastwards towards the west coast of South America, changing the way rain falls around the planet.

As part of this study, anthropologist Benjamin Orlove of the University of California, Davis, reviewed ethnological studies of villages in Peru and Bolivia and found that sightings of the Pleiades are used throughout the high Andes (the Altiplano) to predict the upcoming growing season. A third scientist, John Chiang, Cane's student at the time, was able to link an increase in high-altitude wispy cirrus clouds during June to El Niño events. The clouds dim the brightest star in the Pleiades – the most commonly used forecasting feature – and also obscures the dimmest five of the eleven main stars, the ones on the edge of the constellation, making the Pleiades seem to shrink when viewed with the naked eye. The link between the appearance of this constellation and El Niño explains the success of the Incas and the farmers who have since used the method to forecast rains that will fall months later.

Similar ethnoclimatology studies have been under way in India since the 1970s by scientists at Gujarat Agricultural University, which has a campus in Junagadh, Saurashtra, in the western part of Gujarat state, where drought has been a regular feature. The monsoon season, which extends from June to September, is characterised by irregular, erratic and uneven showers. Local farmers place great importance on accurate predictions of monsoon's onset to work out which particular crops should be planted. Early onset of the rains suits crops with a long growing season, such as groundnut and cotton. When the rains are delayed, the choice of crop is restricted to pulses, pearl millet, castor and one type of groundnut.

A systematic study of traditional forecasting methods was undertaken in 1990. The official Department of Meteorology had predicted normal monsoon for the whole of India. However, that said little about important variations at a local level. To see if traditional methods could say more, university lecturer Purshottambhai Kanani met Devji bhai Jamod, an engine driver in Jetalsar village, and Jadhav bhai Kathiria, a schoolteacher from Alidhra village, both of whom were fascinated by traditional forecasting. They had predicted that monsoon could not possibly arrive before 15 August that year, based on the traditional belief that: 'If there is rain, accompanied with lightning and "roaring of clouds" [mild thunder], on the second day of Jayastha [the Hindu lunar calendar, corresponding to May–June], there will be no rain for the next seventy-two days.'

These words were written some time around the twelfth

century by Bhadli, whose couplets (*Bhadli Vakyas*) have been passed down by word of mouth. They describe ten 'chieftains' (variables) responsible for the development of the 'ethereal embryo' of rain: wind, clouds, lightning, colours of the sky, rumbling, thunder, dew, snow, rainbow and the occurrence of an orb around the moon and the sun.

Just as Bhadli had predicted, there was a downpour seventy-two days later, on 15 August 1990, on Saurashtra, enabling farmers to plant the crops with the shorter growing season. This impressed Kanani so much that he publicised it in the local press and made an appeal to readers to find out more about weather folklore.

Local folk culture relies on the ideas of other astrologers as well as Bhadli, such as Varahmir (*c.* eighth century), Ghagh (*c.* thirteenth century) and Unnad Joshi (*c.* fifteenth century). These magicians made their predictions by observing interactions between wind, clouds and lighting; the flowering and foliage of certain tree species; or the behaviour of birds and animals, whether singing frogs, crying peacocks or climbing snakes.

The beliefs were treated by Kanani as hypotheses and tested over the following eight years. Some do show promise. For example, there is thought to be a link between the wind direction on Holi, a Hindu festival in spring, and the strength of the monsoon that year. Kanani believes that traditional forecasting methods have a role in helping to predict regional variations. Unsurprisingly, not all of his peers are convinced that they should put aside their computers to listen to crying peacocks and so on.

Where magic stops and science starts

Weather prediction methods are examples of how magic can turn into science, once it has been validated by experiment.

One might imagine, therefore, that it is easy to examine the historical record and determine when the influence of magic ends and science starts. It is not, a fact that has been highlighted by the ancient-metallurgy research group of Bradford University. There, Timothy Taylor has looked at magic and the science of metals (metallurgy) down the ages, from the production of crude copper axes to the advanced alloys used to build the space shuttle.

Some depictions of scientific progress suggest that the efforts of societies since the dawn of human prehistory have built a pyramid of knowledge that explained, brick by brick, how the world worked. Once a foundation had been laid on a basic principle, new bricks could be laid upon it. The pyramid model, says Taylor, portrayed a progressive and rational sequence of scientific development in metallurgy, 'through copper, bronze and iron metallurgy, onward and upward to the achievements of the aerospace industry'.

But with his colleague Paul Budd, Taylor points out that this paradigm seems to set metallurgy apart from arts like basketry, flint-knapping and potting, which all evolved 'unscientifically'. They dismiss this distinction and claim that ancient metallurgy was far from scientific and impossible to fathom without reference to the culture of the day. 'Metal-making and magic-making can easily go together.'

In societies where writing was rare or non-existent, metallurgical recipes were committed to memory as a formulaic 'spell'. When these spells were carried out, utility was not necessarily paramount: those able to put on a show of flames, sparks and other pyrotechnics during metal-making may well have earned greater respect because they had a more crowd-pleasing 'ooh factor'. At that time metallurgy was part and parcel of ritual activities carried out by a figure such as a shaman, whose spiritual power was confirmed by scientific and technological power.

Taylor argues that science alone cannot explain key events

in the history of metallurgy, such as the rise of iron over bronze (an alloy of copper and tin) after 750BC. He illustrates this with an example. Imagine if a society with ritual and competitive bronze-making was suddenly overtaken by invaders. At a stroke, the invaders disrupt tin supplies while importing new technology, based on iron, which is glamorised by its warlords and associated spiritual leaders so that a gypsy blacksmith can rearm them more quickly than a Bronze Age chief who had to consult the right ritual channels. 'It is perhaps not the rise of science which accounts for the trajectory of Old World metallurgy, but the power of magic.' In this way, Taylor argues that we must 'put the magic back' into the history of science.

Magic is still at work in this era of New Age medicine, psychic phenomena and paranormal interest. Perhaps this should not come as a surprise, since superstition thrives on uncertainty, which is far from absent in the contemporary world where many things lie beyond our control, from the call of a tax inspector to a diagnosis of cancer.

Cultural anthropologists, Muggles who specialise in magic, have yet to find any society that does not have a long-standing and elaborate system of paranormal beliefs. In this respect, our society is no different from supposedly primitive cultures. Many people in the West rely on astrology and alternative medical treatments. Indeed, the interest in the latter has become so widespread that it now influences the agenda of serious medical research.

Even if we accept that magic has been downgraded in some way with the advancement of civilisation, it is difficult to link this development to the rise of science. Other factors have also made us less dependent on magic, whether it is the rise of literacy or the increasing security of life. Indeed, the link between the rise of science and the demise of magic may be another example of a pattern that appeals to our common sense but is fundamentally false.

The rise of science took a long time to deliver the goods in terms of material benefits. Technology, which has had a huge impact on humanity and ranges from the origins of agriculture to the construction of great cathedrals, pre-dates science. Even after the work of Newton and others, the impact of science on everyday life was far from obvious. The steam engine probably owed more to blacksmiths and old-fashioned ingenuity than any of the lectures on newfangled ideas delivered by the scientists – natural philosophers – of the Royal Society, the UK's venerable academy of sciences.

William Harvey may have presented his theory of the blood's circulation in 1616, but it did not save any lives in the short term. Indeed, sanitation probably did more to improve the general state of health than advances in understanding anatomy. Basic observations that linked cause and effect did more to fight a given disease than basic understanding of the disease agent. Edward Jenner observed that milkmaids rarely contracted smallpox, which led him to use less virulent cowpox as a vaccine. John Snow observed in 1854 that an isolated cholera epidemic in London could be traced to a contaminated water pump in Broad Street.

Today, in some fields of science, there is a clear link between fundamental scientific understanding and quality of life. When scientists find a site in the body where a chemical acts, such as a messenger chemical, it opens up ways to tinker with that signalling process that could be used in treatments. In one sense it is still magic, for few people understand it. The underlying science remains a mystery to the woman on the Clapham Omnibus or the fat man in Number Four Privet Drive. Indeed, many details of the mechanism of action of drugs and other treatments remain mysterious even to scientists themselves.

This type of magic rests on the inability of the brain to comprehend certain processes. However, the human brain can also *create* magic, sorcery and illusions. There are many ways that it distorts the way the world looks and how we remember it. It even plays with time. As we are about to see, the brain is the greatest wizard of all, and its faltering attempts to make sense of the world are at the heart of all magic.

CHAPTER 9

The greatest wizard

'Tis nothing but a Magic Shadow-show,
Played in a Box whose Candle is the Sun,
Round which we Phantom Figures come and go.

Edward Fitzgerald

Real-world magicians of the kind who bask under the bright neon lights of Las Vegas achieve the impossible each and every performance. Yet there is an unspoken understanding between audience and magician that nothing supernatural is involved. Unlike the events that take place under the shivering flames of Hogwarts' floating candles, there is no real magic in Vegas. That is why we refer to the feats of Muggle magicians as illusions, sleight of hand or legerdemain.

Many kinds of conjuring effects are used to dazzle a Vegas audience: making a rabbit appear out of a hat, or a gold coin vanish; swapping places with a glamorous assistant in an enchanted trunk; transforming the colour of a deck of cards; walking through solid objects and linking apparently solid rings; restoring damaged objects to their former glory; carrying out extraordinary feats of memory and recall; turning an assistant into a leopard; moving objects without touching them (telekinesis); feats of extrasensory perception, such as reading the mind of an audience member; and great escapes of the kind once performed by Harry Houdini.

There are many possible rational explanations for the above, which ultimately depend on smoke-and-mirrors. None requires any special access to the spirit world, sorcery, enchantments or hocus-pocus. As if to underline how stage magic works within the laws of nature, some professional magicians, notably James 'The Amazing' Randi, are vociferous critics of those who claim to have tapped some supernatural or psychic ability.

Mysterious appearances of rabbits suggest that the animal was there all the time but concealed, or secretly placed in the hat; the reverse is true for supposed vanishings; a mixture of these strategies is used for transposition and transformation tricks; telekinesis could rely on unseen manipulation, invisible threads or optical illusions; ESP can be staged with props as varied as a deck of identical cards, a stooge in the audience, or a way to view secret information that has been written down by a member of the audience. Above all else, even the cheesiest lounge act mounts elaborate deceptions by exploiting wrinkles in the way the brain perceives the world, stores memories and invests its attention.

Muggles have an extraordinary ability to deceive themselves (just think of Vernon and Petunia's idealised view of their son Dudley). Magicians make cunning use of this to ensure their deceptions are far from obvious, according to Richard Wiseman of the University of Hertfordshire, a psychologist who has done research in this area and is himself a member of the Magic Circle, the premier magical society. Even if a magician told you how she carried out a stunt – which she won't – many, if not most, magicians do not have much in the way of insight into how these techniques of deception work. Wiseman believes that a study of stage illusions may help scientists understand more about the workings of the brain and, in turn, help in the development of new tricks.

A fundamental skill of stage magicians is knowing how to divert attention. They are able to persuade an audience to look in a certain direction, so as to conceal a rabbit in a hat, for example. How the brain invests its attention is the subject of much speculation by scientists. Some, such as Francis Crick, believe that there is a spotlight of attention located in the thalamus, the gateway to the cortex of the brain. Others, such as Semir Zeki of University College London, argue that the brain is a democracy, where the most active brain region – such as the one that has just detected a wand that has been pointed your way – wins the competition for attention.

Close-up magicians who do coin and card tricks are perhaps the most fascinating of all, because their sheer proximity seems to amplify the impact of the magic on bystanders. It was once thought that sleight of hand was all about speed of movement. It turns out, however, that magicians do not rely on the hand being faster than the eye, but on slower – and less obvious – strategies.

To divert and confuse the brains of onlookers they depend on novelty, movement, contrast, body language, voice and gaze. They know that to make an audience gaze at an object, they should stare at it first. Equally, by looking at the audience they know that the audience will look back at them. Magicians are also aware that a good performance has a set rhythm so as to focus the attention of the audience, then relax it. Above all else, they rely on knowing how the mind can be fooled, even if the eye cannot. That is the primary secret of magic.

Hidden trapdoors, invisible wires and sleight of hand have all been discussed endlessly before, as has the broad range of technological tricks deployed in televised magic. I don't intend to go over this ground again. After all, Muggle magicians tend to get upset when the secrets of their complex and skilful art are revealed. Richard Wiseman, like other members of the

Magic Circle, is sworn to secrecy.

I want, rather, to focus on the hidden accomplice that all magicians rely on, the secret sorcerer that is the master of illusions in time and space. This magician, the greatest of all, weighs three pounds, is as big as a cabbage, has the consistency of an avocado, and resides between your ears. She, or he, is often called the most complex known thing in the universe and carries out the extraordinary feat of building the world we experience – a brain reality. This, as we will learn, is not the same as physical reality.

She takes tiny, distorted upside-down images in your eyes and translates them into patterns of nerve cell activity that create an experience based on your surroundings. Even as you read these words, she is playing tricks, warping your sense of time and space. She fools you every minute of every hour of every day of every week. You may think you can distinguish magic from old-fashioned deception. Think again. Since the work of Gestalt psychologists in the early decades of the twentieth century, much evidence has been published on how 'mind reality' is actually out of touch with the real world. The evidence is there before your eyes; you just have to know where to look. To paraphrase Arthur Weasley, we seem to go to any lengths to ignore this magic, even if it is staring us in the face. Once you have seen this magician in action, the skills of the greatest Muggle illusionist do not seem so special after all.

You don't notice much . . .

We are vulnerable to the magician's art of directing our attention to what he wants us to notice. Indeed, we actually take in very little of what we see, despite everything that seems to be going on in our field of view. Our astonishing lack of attention has been vividly illustrated in an experiment by Daniel Simons of Harvard University and Daniel Levin of Kent

State University in Ohio, which reveals our mind-boggling ability to be hoodwinked. People who were walking across a college campus were asked by a stranger for directions. During the conversations, two men carrying a wooden door passed between the stranger and the subjects. Then the subjects were asked if they had noticed anything change after the door went by. Half of those tested failed to notice that the stranger had been replaced by a man who was of different height, of different build, with short hair of different style and who sounded different. He was also wearing different clothes.

Despite the fact that the subjects had talked to the stranger for a full minute before the swap, half of them did not detect that, after the passing of the door, they had ended up speaking to a different person. This phenomenon, which psychologists call change blindness, highlights how we see far less than we think we do. The brain seems to extract key and salient details and fills in the gaps from memory to free up valuable computing power for other neural processes. To depict the world, the mind functions like a painter: it identifies the essential features of an image and discards redundant information.

Working with Christopher Chabris, Simons played a videotape of a basketball game and asked his subjects to count the passes made by one of the teams. Around 50 per cent failed to spot a woman dressed in a gorilla suit who walked slowly across the scene for five seconds, even though this hairy interloper had passed between the players. However, if subjects were simply asked to view the tape, they noticed the gorilla easily. Some of them refused to accept they were looking at the same tape and thought that it was a different version of the video, one that had been edited to include the ape. Richard Wiseman recently repeated this experiment before a live audience in London (as part of his *Theatre of Science*, performed with the author Simon Singh), and found that only 10 per cent of the 400 or so people who saw the show managed to spot the gorilla.

For all our experience of a rich visual world, it seems that in any given situation we take in no more than a few new facts, add them in a few stored images and beliefs, and produce a seamless whole in which it is impossible to tell what was real and what remembered. Given that we have a less than complete picture of the world at any one time, it is no wonder that stage magic continues to thrive.

. . . and the little you notice is distorted

To make something disappear, you do not need a Conjuctivitis Curse or a cry of 'Deletrius!' Every time you open your eyes your mind makes something that is not there disappear. By the 'something that is not there' I mean the 'blind spot'. The information we receive from our eyes is incomplete: there is a spot on the light-sensitive retina at the back of the eye where light is not detected because it is where the optical nerve sends visual information to the brain.

Even Harry Potter's toad-green eyes have a blind spot. But we only notice it in certain circumstances. A familiar children's puzzle uses the phenomenon to make one of two dots on a piece of paper 'disappear' as the paper is moved closer to the eye. Charles II occasionally pretended to decapitate his courtiers when gazing at their portraits by closing one eye and coinciding the 'blind spot' in the other eye with their heads. It would not surprise me if Harry Potter used the same technique to decapitate ghostly Professor Binns, as he drones on about the history of magic, to alleviate the boredom or simulate Nearly-Headless Nick.

The blind spot has been studied by Richard Gregory of Bristol University and Vilayanur Ramachandran of the University of California, San Diego. They began by asking why, when every eye has a blind spot, we do not see a large black dot in our field of vision. Calculations suggest that the spot should be prominent, around the size of six moons in the sky. In terms of this page, the blind spot is around four or five words away from the word you are reading and about the size of a word.

Various ideas had been put forward to explain how the brain repairs this image (and without so much as a single cry of 'Reparo!'). Either we do not notice the blind spot, or else the brain 'fills it in' by extrapolating from visual information picked up from around the edges of the spot. Gregory and Ramachandran believe that the latter is the case, after experiments in which they showed computer-generated video patterns to volunteers who had artificial blind spots temporarily induced in the eye.

The team compared what they showed the volunteers with what the volunteers reported seeing. Their technique revealed how much the brain creates to fill the blind region. The brain abhors a vacuum, and has to complete an image. For instance, if an image of a straight line passes through the blind spot, the brain deduces from the retinal representation of a broken line that the line is in fact continuous and extends the line across the gap.

There are other forms of blindness we all suffer from without realising it. For example, if you concentrate on a swirling pattern of dots superimposed on a few stationary dots, the latter wink in and out. This 'motion-induced' blindness has been studied by Yoram Bonneh of the Smith-Kettlewell Eye Research Institute, San Francisco. The dots are erased in the mind, not from the image: presented with contrasting stimuli, the brain shifts into 'winner takes all' mode, sacrificing its attention to some objects in the visual field. Bonneh speculated that the phenomenon could occur in real-world situations, such as night driving. When a motorist is presented with many moving head and tail lights, the relatively stationary lights of the car in front of her could in principle disappear.

Motion-induced blindness also reproduces some of the effects of the disorder simultanagnosia, where patients – often with brain damage – are unable to perceive more than one object at a time. One extreme example appears in a book by the Russian neurologist Alexander Luria, *The Man with a Shattered World*, which contains notes made by a soldier wounded during the Battle of Smolensk in the Second World War: 'Just then I happened to look down and I shuddered . . . My hands and feet had disappeared. What could have happened to them?' Once again, it is possible for the brain to make things vanish, as if by magic. As well as ignoring reality, the brain can also create its own reality.

You can see movement where there is none . . .

Some art can be quite moving without being emotional. *Enigma*, a painting in black, blue and white by Isia Leviant, has a curious effect on the observer. Even though the image is static, one can see a slight swirling motion within its solid rings. This is a remarkable example of how the brain can summon up phantoms. Semir Zeki has studied this motion effect, and examined with colleagues John Watson and Richard Frackowiak what *Enigma* does to the brain of an observer. In particular, they wanted to see whether there was a shift in the pattern of brain activity that could be linked with the swirling motion set up by the painting.

Zeki's team had already used a brain scanner to discover that one region in the thin rind on the back of the brain, called V5, is used by the mind to analyse moving visual stimuli. The same region – and some surrounding areas – are weakly stimulated when people look at a black-and-white version of *Enigma*, suggesting that it is activity in this small region that produces the sensation of circular motion. 'For some reason, that static picture causes activity in V5, which is why you see motion when there is no objective motion there,' says Zeki. Although he and his colleagues now know what *Enigma* is doing to the brain, they do not understand why the painting's pattern of concentric rings tricks the brain into creating motion where there is none in the real world. The effect is, in fact, somewhat magical.

. . . and stop the movement that shows the passing of time

Have you ever glanced at a clock and felt that the second hand takes longer than a second to move? Perhaps you halted the second hand by shouting 'Impedimenta!' No such magic is actually necessary. The brain can easily stretch time without spells because of the way it edits reality. When our eyes move to focus on a clock face, the image we see should be blurred. But the brain casts a spell to ensure that everything looks crisp by cutting off our vision for the period the eyes are in transit. To stop you becoming confused, it then adds on the time taken to move the eyes to the impression of the next stable image. Intriguingly, it makes the image appear *earlier* than it actually did. That way, there is no gap in consciousness.

Our brain can stretch our perception of what we see backward in time to 1/20 of a second before we started to move our eyes, rather than starting the perception when our eyes reach actually where they intend to look, according to a study by Kielan Yarrow and John Rothwell of the Institute of Neurology in London. If the second hand had moved just before your eyes stopped moving, you would then feel that more than a second had passed before the hand's next movement. And indeed, thanks to the time warp ability of the brain, that is the case.

The length of the illusion of chronostasis (the stopped clock) is linked to the size of the eye movement preceding it, and suggests that the illusion rests upon an assumption made by the brain – that the target of an eye movement is motionless even when we cannot actually see it. We have the impression of a seamless and full sensory experience unfolding as things happen, but a closer inspection reveals this to be the result of assumptions made by the brain to smooth over ambiguous situations. We become aware of these assumptions only when clock watching.

You may also reasonably link the sound of a 'tick' with the precise moment that a clock's second hand moves. But sound and vision are processed by different parts of the brain and sometimes they are not quite knitted together. This was revealed by a series of experiments to show what we mean by 'now' that were conducted by Jim Stone of Sheffield University in England. If a firecracker goes off at arm's length, the time that it takes a person to become aware of the bang is different from that of becoming aware of the flash because each sense is handled by different circuitry in the brain. After 1,000 'flash bang' trials on seventeen people, Stone discovered that there is a wide difference in the time taken to process sound as opposed to vision. Some people perceived light up to 21 milliseconds before sound, suggesting they could process light more quickly. However, most people hear sound first: for one it took up to 150 milliseconds – about one seventh of a second – to register light after hearing the sound. (In this particular case, the person actually experiences a slight lag between hearing someone's voice and seeing their lips move.) 'We found that the "now" is different between individuals, but it is very stable within each individual,' Stone tells me.

Some very eminent Muggles have become confused as a result. Individual differences in sound and light perception may even explain why David Kinnebrook, a new assistant to the Astronomer Royal, Nevil Maskelyne, was fired in 1794. Kinnebrook was asked to record observations of stars and their locations by the 'eye and ear method', which relied on listening to ticks of a clock and viewing the time at which a given star drifts between two lines. Maskelyne checked Kinnebrook's work but came to different conclusions, from which he deduced that Kinnebrook was half a second slower. Kinnebrook's attempts to improve failed, and he was sacked. Two decades later, the influential German astronomer Friedrich Bessel concluded that the differences in observations were due to differences between these scientists – and the wrinkled wizards in their heads – and not to inferior work.

The colour of motion

We can uncouple more than just sound and vision. The binding between motion and colour can come unstuck, too. Imagine a metronome in which the bar changes from red to green and back again as it swings to and fro. It turns out that at certain speeds the colours seem to move the opposite way to reality. In a study of nine volunteers, using moving coloured squares rather than a metronome, Semir Zeki found that if his subjects concentrated on the colour of upward moving red squares, they would associate this colour with the subsequent downward movement of green squares that appeared on the screen 8/100 of a second later. The reason, once again, is that different brain areas work at different paces.

By the standards of signals that crackle through the brain, eight-hundredths of a second is a long time. The mix-up occurred because we can process the colour of something faster than its movement, which explains why the subjects 'bind' the red squares to the later movement by green squares. The discovery suggests that the brain must work harder to process information about motion than colour. This is surprising because the motion-processing part of the brain has a head start when it comes to dealing with information from the eyes. Evidence has shown that motion signals arrive at the appropriate part of the brain before colour signals.

These kinds of experiments have profound implications. In

the old days, neurologists believed that the brain was organised in a hierarchical way: different regions reported to a 'master area'. But nothing approximating this dictator has been found. Then it was thought that all of the separate areas all talked to a 'synchroniser', which integrates the information on motion, colour and so on, presenting the mind with a coordinated whole. 'But there is no synchroniser, either,' says Zeki.

Instead, the brain seems to work in a democratic and interactive fashion. When we glance at something, the light-sensitive cells in the retina send visual information to all regions of the visual cortex, including the colour-processing centre, V4, and the motion centre, V5. Given enough time, these different areas are able to 'talk' to one another to bind the various streams of information into a unified whole in the consciousness. But when events occur in faster than 100 milliseconds – 1/10 of a second – the speed limitations within which each part of the visual mechanism become significant, and our unified impression of the world begins to fall apart.

Sometimes the brain resolves a conflict between the senses by taking sides, as in the example of ventriloquism, where the origin of the sound made by the ventriloquist and the lip movements of his dummy do not agree. Our vision is better at pinpointing location than ears locate sound, and thus lip movement dominates over the voice of the ventriloquist when the brain integrates the two: that is why the sound appears to come from the dummy's mouth.

What we see is what we hear

The brain's democratic nature can be seen in other revealing experiments. In one study by Steven Hillyard of the University of California, San Diego, and colleagues, subjects were instructed to say whether a dim, obscured light appeared soon after a sound was made. Hillyard found that the light was detected more accurately when it appeared on the same side of the person as the sound: what people hear influences what they see. 'Our results suggest that you will see an object or event more clearly if it makes a sound before you see it,' he said. One can imagine a magician using such a trick to divert the attention of the audience.

However, sound can do more than merely guide vision. Experiments by Ladan Shams of the California Institute of Technology, Pasadena, showed that sound can actually change vision so that what you see is what you hear: when a single flash is accompanied by several beeps, we wrongly perceive what we see as several flashes. This ability to create light where there is none is reminiscent of the Hand of Glory, glimpsed by Harry Potter in Borgin and Burkes, which only gives candlelight to the holder.

The experiment undermines the idea that the senses work independently, and that the brain processes information from each sense separately, putting them together to build up a picture of events around us. In fact, our vision profits from signals from other senses. This mixing of faculties boosts the bang per buck that the body obtains from the senses.

Memories are not made of this

Memory Charms are used by the Ministry of Magic as they struggle to keep the wizarding world secret from Muggles. Indeed, this form of enchantment is so routine that they employ Obliviators to do the job. However, a wide range of research suggests that it can't be a particularly demanding profession, given the fickle and creative nature of Muggle memory.

We have less than total recall, a finding emphasised again and again since the first careful studies were carried out more than a century ago by Hermann Ebbinghaus, the German experimental psychologist who published the pioneering work *Memory: A Contribution to Experimental Psychology* in 1885 and developed his 'curve of forgetting' that revealed how, within an hour of learning, more than half of the information learned was forgotten. Given how our identity depends on our memories, can we ever be sure of who we really are if we can't rely on what we remember?

Eyewitnesses are not able to build up a reliable picture of a suspect in a crime from a fleeting glimpse. People typically recalled a clean-shaven man as having a moustache, straight hair as curly and so on. Hypnosis, in turn, can plant false memories. Memories that have faded are vulnerable to distortions. The power of pure suggestion has made people believe that they were born left-handed, how they got lost as a toddler, spilled punch at a wedding and broke a window. Television advertisements depicting nostalgic childhood scenes are also able to fabricate memories: in one US study, adults 'remembered' drinking Stewart's root beer from bottles in their youth, although the bottles had been in production for just a decade, before which drinks were available only from a soda fountain.

'A flimsy curtain separates memory from imagination,' says Elizabeth Loftus, an expert in the field. Among her many research projects, she studied the consumer version of false-memory syndrome in 120 subjects at the University of Washington in Seattle, with Jacquie Pickrell. They revealed the malleability of memory using a fake advertisement showing 'visitors' to Disneyland meeting Bugs Bunny. Cartoon aficionados would know, of course, that Mr Bunny was never a Disney character but a creation of Warner Bros. The two had never appeared together and yet a significant number of people vividly recalled that magical moment. About one-third of the subjects who were presented the advertisement later said they remembered or knew that they had indeed met Bugs Bunny in this impossible setting.

'The frightening thing about this study is that it suggests

how easily a false memory can be created,' remarks Pickrell. 'It's not only people who go to a therapist who might implant a false memory or those who witness an accident and whose memory can be distorted who can have a false memory. Memory is very vulnerable and malleable. People are not always aware of the choices they make. This study shows the power of subtle association changes on memory.'

The way our memories are shaped by the environment of ideas, whether the attitude of the Church towards witchcraft or Hollywood blockbusters about demonic happenings, was underlined in another study by Loftus. She notes that the number of people who report demonic possessions increases in the wake of movies and television programmes that deal with exorcism. The publication of the book *The Exorcist* in 1971 and of the film version at the end of 1973 generated a mini-epidemic of demands for exorcisms. 'Quite a number of people who watch these exorcism films will be affected and develop symptoms of hysteria,' says Loftus, who with Giuliana Mazzoni and Irving Kirsch conducted a study to find out how to induce 'implausible autobiographic events'.

Their research on 200 students demonstrated how a significant minority – nearly one-fifth – of those who previously stated that demonic possession was not very plausible could be persuaded to change their minds. The researchers tampered with belief and memory to introduce the idea in a few simple steps that the subjects thought they may have witnessed a possession in childhood: subjects were shown articles on demonic possession that said the phenomenon was more common than thought; they were asked to list their fears; and finally the subjects were told that witnessing a possession during childhood caused those fears.

'We are looking at the first steps on the path down to creating a false memory,' says Loftus. 'There is controversy about whether you can plant memories about events that are unlikely to happen. As humans we are capable of developing memories of ideas that other people think occurred. Just being exposed to credible information can lead you down this path. This shows why people watching *Oprah* or those in group therapy believe these kinds of things happened to them. People borrow memories from others and adopt them as their own experiences. It is part of the normal process of memory.'

Loftus's research reinforces the idea that therapists need to be prudent when they help patients recover 'lost' memories of early traumatic events. Subtle suggestions and prodding for greater elaboration could effectively plant false memories, whether of UFO abductions or a serious injury suffered in a past life or a vivid eyewitness account of satanic rituals. Indeed, given current Pottermania, it would be reasonable to expect encounters with imaginary witches and wizards to be on the increase.

Bewitched dreams

'The presence was of a demonic nature, purest evil, out to possess my soul . . . I find this utterly terrifying, beyond anything I can imagine experiencing in the real world because it is so contrary to reality and yet feels entirely authentic.'

This response to a survey of thousands of people worldwide is testament to how, even without hallucinogens, the mind can summon up strange experiences that in days gone by could be blamed on witchcraft. Over a lifetime, many people will fall into a curious penumbra of consciousness in which they sense the presence of a nearby threatening evil. I experienced it myself as a teenager. The details are embarrassing to recall. In the small hours I awoke at home and saw a cowboy with a gun. I wanted to call out, but I was paralysed by fear. A weight was bearing down on my chest, and I struggled to breathe.

I lay in bed for what seemed like an age before I finally realised that the 'cowboy' was actually the shadow of a lampshade. You might think that I laughed the mistake off and went back to sleep. In fact, I was still shaken. I had never experienced such primal fear. I felt sheepish, if not a bit foolish. Only when I was older and (supposedly) wiser did I learn that I had suffered what is called sleep paralysis – a kind of breakdown between brain and body that takes place on the fringes of sleep, either when falling asleep or awakening.

The details of such episodes vary from person to person. Some hear vague rustling sounds, indistinct voices and demonic gibberish, while others experience hallucinations of humans, animals and supernatural creatures. One common element is a striking inability to move or to speak, or the sensation of a weight on the chest. Also typical are feelings of rising off the bed, flying, or hurtling through spiral tunnels. Not surprisingly, these bizarre experiences are accompanied by fear and terror. Usually, after a minute or two, the spell is broken.

Sleep paralysis was once thought rare. But recent studies by Kazuhiko Fukuda, a professor at Fukushima University in Japan, suggest that it may strike between 40 and 60 per cent of all people at least once. About 4 per cent of the population experiences sleep paralysis on a regular basis. Fukuda also found that while sleep paralysis was equally prevalent in Japan and Canada, citizens of the latter often wrote off the experience as a dream because there was no common expression for the condition. In Japan, in contrast, it is known by the name *kanashibari*. There is a good chance that Harry Potter, sleeping in his four-poster bed in the tower dormitory, may have also experienced the effect.

The phenomenon has been recorded down the ages. *The Bewitched Groom* by the German painter and printmaker Hans Baldung Grien is typical of Renaissance imagery in which artists depict sleepers tormented by animals, goblins or witches. A vivid account of the hallucinations caused by sleep paralysis appears in Guy de Maupassant's 'Le Horla', a short story where a frightening presence – the Horla – plays a central role. The phenomenon may also have inspired the 1781 Henry Fuseli painting *The Nightmare*, which depicts a goblin sitting on the stomach of a sleeping woman.

This goblin is well travelled. The word 'incubus', for a demon that stretches out upon sleeping people, is derived from the Latin *incubare*, 'to lie upon'. In Newfoundland, sleep paralysis is called 'old hag' because it is often linked with visions of an ugly old woman squatting on the chest of a paralysed sleeper, sometimes throttling them. (Harry encountered one – who ate raw liver – in the Leaky Cauldron.) The Chinese refer to 'gui ya', or ghost pressure, when a phantom sits on and assaults sleepers. Parts of Germany referred to *Hexendrücken* for similar reasons. Far away, in the West Indies, there was *kokma*, when a ghost baby bounced on the sleeper's chest and attacked the throat. In ancient Japan, a giant devil was blamed. The experience probably gave us the term nightmare, given how many languages refer to malevolent spirits of this sort with variants of the term 'mare' (Czech, *muere*; Polish, *zmora*; Russian, *kikimora*; French, *cauchemar*; Old English, *maere*; Old Norse, *mara*; and so on).

One expert in the field, Allan Cheyne, thinks it is significant that Harry Potter's wizard background is first revealed to him by Hagrid. '"Hagrid" is a term used in Newfoundland to describe the appearance of someone who has been "hagged" or hag-ridden the previous night,' he says. 'It is sometimes also thought to be a corruption of the word "haggard" (or perhaps haggard is a corruption of hag-rid).' The Oxford English Dictionary also vaguely alludes to the connection between the word 'haggard' and 'hagged'. (As a curious aside, the dictionary also points out that the term 'haggard' refers to an owl that has its adult plumage.) Finally, the dictionary provides this titbit from Thomas Hardy's *Mayor of Casterbridge*: 'When she had not slept she did not quaintly tell the servants next morning that she had been "hagrid".' Thus, as Cheyne points out, 'The term "hagrid" does not seem to be original to Newfoundlanders.'

To find out how people attempt to make sense of these unusual, frightening and hallucinoid sleep experiences, Cheyne has studied the responses to the Waterloo Unusual Sleep Experiences Survey on nearly 11,000 cases from around the world, gathered at the University of Waterloo in Canada.

The responses to his survey are chilling. There are reports of spectral figures: 'I saw a black humanoid shadow move over to the ceiling above my head, and then it seemed to glide down on top of me.' This sounds rather like the Lethifold, a creature resembling a black cloak that envelops its sleeping victims. Others summon up one of Harry's worst nightmares, a Dementor; one respondent described a presence 'trying to suck out what I thought was my soul'. Another remarked: 'It wants to take my soul or mind or remove me from my body.'

Whatever the cowled figure was, the respondent was 'absolutely sure it was supernatural and evil'. Some people hear a presence, sometimes even detecting a smatter of Parseltongue: 'I awoke to find a half snake/half human thing shouting gibberish in my ears.' Others have even been grabbed by a ghost: 'Once the presence was a dark shadowy evil figure and once a white mist, which called my name and touched my shoulder.'

People made sense of them by drawing on what seemed plausible at the time, so the way these vivid experiences are interpreted depends greatly on the culture, says Cheyne. Hundreds of years ago, these interpretations included witches forcibly taking victims for a ride on a broomstick. Even with the rise of popular science in the nineteenth century the experiences were not rationalised as something to do with the brain but instead stimulated a new set of explanations. Today people are more likely to report alien abductions if they have seen movies or read books discussing the topic. Cheyne says people may also report seeing Darth Vader or Freddy Kruger.

Our culture is brimming with stories of outer space and UFOs, and these have become the most readily available answer to a mind struggling to make sense of sleep paralysis. In his 1994 book *Abduction*, for example, Harvard psychiatrist John Mack claimed 'several hundred thousand to several million Americans may have had abduction or abduction-related experiences'. Given the prevalence of sleep paralysis, that extraordinary estimate actually seems plausible.

'A sensed presence, vague gibberish spoken in one's ear, shadowy creatures moving about the room, a strange immobility, a crushing pressure and painful sensations in various parts of the body – these are compatible not just with an assault by a primitive demon but also with probing by alien experimenters,' Cheyne says. 'And the sensations of floating and flying account for the reports of levitation and transport to alien vessels.'

The sense of another's presence, fear, and auditory and visual hallucinations are believed by Cheyne to arise from the brain's entering a hypervigilant state, which reflects events in the midbrain, the normal function of which is to resolve ambiguities inherent in threats. This neural machinery helps sort the animate from the inanimate but, during sleep paralysis, could instead endow a shadow or vague shape with the sense of something living, a sacred or demonic presence. 'Such a numinous sense of otherness may constitute a primordial core consciousness of the animate and sentient in the world around us,' says Cheyne. The source of the fear that this 'numinous sense of otherness' triggers also lies in the brain. Sleep paralysis sees the superposition on wakefulness of dreaming, when the seat of the emotions, the brain's limbic system, plays a prominent role.

The netherworld between sleep and consciousness where sleep paralysis thrives is described by Emmanuel Mignot, director of the Center for Narcolepsy at Stanford University Medical School and, with his two sons, a Harry Potter fan. Sleep paralysis seems to occur when the body is in REM – dreamsleep. During REM sleep – so named because of the rapid eye movement that takes place – the brain essentially disconnects itself from the body. Animal experiments have shown that this shutdown occurs to prevent sleepers from acting out their dreams and nightmares, or cats running about in their sleep as they chase mice.

Because the brain's motor centres remain off line as they slip into this twilight form of consciousness, sufferers are paralysed to the point that they lack even automatic reflexes, like kicking when the knee is tapped. This may also explain the breathing difficulties, or pressure on the chest. Mignot speculates that in the past people who often found themselves in this strange region between dreams and wakefulness may have been more likely to become shamans who claimed to be able to enter the spirit world. The link between sleep paralysis and conscious dreaming (what scientists call a hypnagogic state, entered before one is fully asleep) may even suggest one possible treatment: one could drink a dreamless sleep potion of the kind familiar to Harry Potter and, in that way, avoid the REM sleep state where sleep paralysis is born.

Ghosts that haunt the mind

The brain is a knowledge-gathering machine and, to work effectively, it must focus on what evolution has taught it to be important. It carries out endless abstractions to form concepts, whether of a car, a line or falling in love. However, witches and fairies also lurk within it – along with more ghosts than can be seen at a Deathday party. These have been revealed by studies of hallucinations suffered by people who are untouched by brain damage or psychosis but are losing their sight.

They are called Charles Bonnet hallucinations, after the Swiss naturalist who reported his grandfather's strange experiences and later went on to suffer the hallucinations himself. Dominic ffytche of the Institute of Psychiatry in London has studied many of these patients. 'The hallucinations tend to be brief – lasting a few seconds or minutes – reappearing after hours, days, months or even years. The experiences may be frightening but, with time, most sufferers recognise them as hallucinations and learn to ignore them. Some even find comfort in their bizarre, amusing forms and intricate visual detail – detail which they are unable to see in real life.'

Strikingly, ffytche found patterns in these visions. Rather than witnessing anything and everything, the patients report apparitions that usually fall into a handful of categories, including distorted faces, costumed figures and other bewildering sightings. 'I'm sure fairies and witches all relate in some respect to these disembodied hallucinations,' he says.

Ghosts are among the categories. One patient described how a friend working in front of a tall privet hedge suddenly disappeared, as if he had put on a cloak of invisibility: 'There was an orange peaked cap bobbing around in front of the hedge and floating in space by its own devices.' There are also ghouls. The disembodied or distorted face of a stranger with staring eyes and prominent teeth is seen by about half of all patients, sometimes only in an outline, cartoon-type form. Ffytche calls this prosopometamorphosia. The faces 'are often described as being grotesque, or like gargoyles'.

In the hallucinations, objects or people often appeared much smaller than usual, but sometimes were much larger (micropsia and macropsia). Could these distortions be linked to sightings of the little people and giants? Other patients suffer polyopia, in which one object in the field of view is multiplied in rows or columns, creating multiple heads, a lawn of birds or a wall of coffee mugs. One patient described 'two half heads joined like Janus', which sounds rather like what Harry Potter encountered at the back of Quirrell's head, where he gazed at a second face, the most terrible he had ever seen. Another patient seemed to catch a glimpse of the dreaded Dementors: some of the faces conjured up by his failing eyesight have blank eye sockets with 'expressions [that] are all evil and malevolent'.

When figures were reported in the hallucinations, these phantoms were typically small, wore period costume, and moved in a realistic way. Overall, 40 per cent of sufferers saw figures in costume. 'These could be Edwardian costume, knights in shining armour, military uniforms, Napoleonic uniforms and First World War uniforms,' says ffytche. 'They often wear hats or helmets.'

Some hallucinations consist of geometrical patterns that bear a resemblance to those painted in prehistoric Europe. Today, similar visions are seen by Namibian Ju'hoansi tribespeople during a nightlong dance associated with rhythmic clapping and chanting, when the tribe supposedly adopts the power of animal spirits and uses them to heal. As well as random patterns and geometrical forms, ffytche's patients also saw serene landscapes and vortices. Near-death experiences could be activating the same brain areas that produce these hallucinations, he speculates.

Reassuringly for the patients, these figments are not a sign of madness. These phantoms appear when vision deteriorates to a certain point, usually as a result of eye diseases such as age-related macular degeneration, glaucoma and retinitis pigmentosa. As the brain is starved of sufficient information from the eyes, it compensates with abnormally increased activity and conjures up hallucinations from the random firing of nerve cells. 'You get the same phenomenon in patients who have had both eyes removed,' he says. 'When there is no information coming in, the brain is idle, cells are firing away and producing these stereotyped categories of hallucination.' And he believes that something similar may result from sleep paralysis.

Using a brain scanner, ffytche has observed that the thin rind on the back of the brain, where vision is processed, is active during these hallucinations, just as it is when a sighted person opens his eyes. The visions fall into a series of specific categories because the brain uses a systematic approach to make sense of degraded information from the eyes.

These are the rules that the brain uses to instil meaning into visual information. But it is not the residual degraded input from the eyes that is important, says ffytche, but the absence or loss of visual input that triggers spontaneous activity in specialised visual areas of the brain to create these phantoms.

Ffytche believes that by studying the hallucinations, scientists will gain a great deal of insight into how the brain processes vision, and he has already reached some conclusions about the origins of the apparitions. The location of the brain activity during hallucinations matches what is known about how the brain processes vision, so colour hallucinations saw crackles in the part of the cortex used to handle colour, face hallucinations saw more activity in the part of the cortex that handles faces and so on. But where did the gargoyles come from? Ffytche points out that one part of the brain – the lateral occipital region – alerts us to the possibility that what we are looking at might be a face. This region detects a face's component features – the eyes, nose, lips and chin, for example – but does not register where these features are. It does not care, for example, if a chin is located on the forehead or the eyes under the nose. 'Our results showed that it was this face-feature detector which caused the gargoyle-like hallucinations – its indifference to the position of each feature leads to the characteristic distortions of the gargoyle and the overemphasis of face features, the prominent staring eyes.'

The recurring association of figure and garden hallucinations seems to relate to how the area of the brain that processes landscapes, gardens and vistas lies next to the one responsible for processing figures and objects. Activity in one is likely to spill over to the other, leading to a spurious association of the two hallucination categories. The abundance of hats could reflect how cells in the temporal lobe in the brain, which encode elongated shapes, are firing. But ffytche admits that it remains a mystery why the figures should so often be costumed and sporting flamboyant hats, looking like they just walked out of Hogwarts.

The Mirror of Erised

One of the most haunting moments in the first Harry Potter book comes when our hero gazes into the Mirror of Erised. At that moment, Harry found himself looking at his family for the first time in his fully conscious life. The mirror has a special property that is central to the plot. It can read your deepest and most desperate desire, your most heartfelt dream. (In the case of Dumbledore, that happens to be a pair of thick woollen socks.)

Could the brain ever be fooled into believing a looking-glass reality? An Alice in Wonderland syndrome, one that makes otherwise sensible people believe that reflections in a mirror are real, was reported by Vilayanur Ramachandran of the University of California at San Diego. With Steven Hillyer and Eric Altschuler, he explained how patients who had suffered a stroke could sometimes suffer the looking-glass syndrome – mirror agnosia – in which they would bump their hands against the mirror glass trying to reach an object 'inside' the mirror. Lewis Carroll's story of Alice stepping through the looking glass into another world may have been inspired in part by this bizarre brain disorder.

But how did Harry see his parents? One way to conjure up a vision of the deceased relies on brute force technology: scientists could presumably use a brain scanner and, with a profound understanding of brain activity that we still do not possess, deduce Harry's deepest desire, then project it on the mirror, in reality a high-technology screen. However, the above discussion shows that there is another possibility. Perhaps Harry's own brain generates the vision of his parents.

Bereavement hallucinations are a well-recognised feature of

the grief that follows the loss of a loved one, says ffytche. Studies have estimated that around 10 per cent of grieving people believe that they have caught a glimpse of their lost spouse, just as the lovelorn believe they have seen the object of their desire when in fact it is a stranger who walks, talks or dresses the same way. These bereavement hallucinations occur without eye disease and relate, presumably to 'higher cognitive functions' of the brain. Albus Dumbledore refers to how the dead we have loved never truly leave us and, in this sense, he was right.

Another story lies behind the mirror. J.K. Rowling was greatly affected by the death of her mother from multiple sclerosis in 1990, aged only forty-five. She died before the extraordinary success of the Potter series would propel her daughter to international stardom. By creating the magical Mirror of Erised ('desire' spelled backwards) in the first Harry Potter novel, Rowling movingly commemorated her mother and the ache left by her loss.

The mind's Imperius Curse

The wizard in our heads may be powerful but he takes pity on his Muggle host and is at least kind enough to create the illusion that we Muggles are in control of our actions. Take the simple act of turning on a light, for example. To do this, you would expect the part of the brain that makes a conscious decision to go into action *before* the part of the brain that actually controls the movement of the finger that flicks the switch.

Wrong. Two decades ago, scientists made the puzzling discovery that when a person is asked to lift his finger, he becomes aware of the urge to move some 300 milliseconds *after* brain activity to trigger the movement. This sounds like the wizard between our ears is casting something like the

'Imperius Curse', one of the 'Unforgivable Curses', a spell that makes someone do anything the caster of that spell wishes.

The strange effect has been investigated by Patrick Haggard and colleagues at University College London by looking at what happens in the brain when subjects decide to move a finger to trigger the noise of a 'beep'. To create the illusion of control, it turns out that the brain plays a trick: when an individual is asked to say when she thought she moved her finger, she estimates that the movement took place *later* than it actually did. Similarly, the beep was perceived by her to happen *earlier*.

Free will seems to have the effect of making us think that cause (pushing the button) and effect (the beep) happen more closely together than they really do. As a consequence, 'The brain binds together our actions and their effects to produce a coherent conscious experience,' says Haggard. This 'binding' reflects our feeling of being in control.

The team took this study one step further and used a technique called transcranial magnetic stimulation, in which electromagnetic coils are applied to the skull to activate the motor cortex of the brain, which controls movement. This made the subjects' fingers move without their cooperation. When treated like puppets, subjects then perceived that the involuntary movement occurred earlier and the beep later than when they intended to push the button.

This suggests that, without free will, the increase in perceived time between the involuntary movement and the beep for the movement may reflect the brain's attempt to separate two events that cannot plausibly be related. Fascinatingly, the trick used by the brain to create the illusion of control is sometimes lost, leading to the feeling that one has been subject to some kind of mind control curse, or is a robot carrying out a program, comments Haggard's colleague Chris Frith.

Some schizophrenics complain that if they reach for a glass, for example, 'It is not me that is doing it but an alien force.'

This suggests that they may be aware of picking up the glass before becoming aware that they wanted to do so. As Haggard's team has shown, the time at which we think our actions occur is a construction or trick of the brain to create the impression that we are in control. Schizophrenic individuals may in fact be more in touch with reality than normal people.

The hocus-pocus of consciousness

The above discussion should have persuaded you that it does not take much to create odd illusions in which the reality manufactured by the brain does not match that of the outside world. To make something disappear, you do not necessarily need a wizard to cast a Memory Charm by muttering 'Obliviate' or whatever. The wrinkled wizard in your head already does famously well when it comes to distorting memory and perceptions or creating the illusion of free will.

The brain not only misses key details of the world but can also generate illusions, visions and phantoms. Sleep paralysis is accompanied by a legion of odd characters and creatures, such as demons, hags and witches. They are so real that one can see, hear and even feel these apparitions.

Hallucinations caused by fading vision are able to summon other peculiar creatures, make things invisible, or litter our field of view with birds. And for the bereaved, it is even possible to conjure up the ghost of someone they've lost, a neural testament to the power of love. Now imagine the impact on the Muggle mind of the strange phenomena and sights that it has been exposed to over the millennia. From them, Muggles have thought themselves in the presence of dragons, werewolves, giants and much more. Mythology is born of the mind's struggle to understand its surroundings.

CHAPTER 10

There be dragons. Really

Science must begin with myths.

Sir Karl Popper

There are many references to the most famous of all magical beasts throughout the Harry Potter books. Dragons guard the high-security vaults of Gringotts Bank. A temperamental black and lizard-like Hungarian horn-tail faces Harry at the Triwizard Tournament. We learn that there are various species of the fire-breathers, such as the Swedish Short-Snout, which is horned and blue-grey; the Chinese Fireball, red with fine gold spikes; and the Antipodean Opaleye, which has iridescent scales. Finally, of course, there is the well-known reference to the beasts in the Hogwarts motto: *Draco dormiens nunquam titillandus*. As anyone with a shred of common sense already knows, never tickle a sleeping dragon.

An onus has been placed on wizarding governing bodies to cover up the tracks of these creatures since 1750, when a clause was added to the International Code of Wizarding Secrecy. Spells are put on Muggles to make them forget they have seen dragons, even a little one like Norbert the Norwegian Ridgeback. These efforts have not always succeeded, however.

Newt Scamander, author of *Fantastic Beasts*, details the Ilfracombe incident of 1932, when a Welsh Green dragon terrorised British sunbathers, but concludes that wizards can congratulate themselves on a job well done in covering up the matter.

I beg to differ, for I have found evidence of dragon sightings that date back many hundreds of years. In Britain, they are part and parcel of legend and folklore. A 'fiery dragon' flitted between two hills near Cadbury, Devon. A 'devouring dragon' with a taste for milk was killed by Sir Maurice de Berkeley in Bisterine, Hampshire. Another had his head chopped off while still another slept in the sun, with his 'scales ruffed up'. There are reports of when a black dragon clashed with a 'reddish and spotted' dragon on 25 September 1449, near Little Cornard in Suffolk. A five-headed variety caused havoc in Christchurch, Dorset. Nine maidens were consumed by one Scottish beast. Other dragons went by the name of the Knucker, Laidley Worm or, in Wales, the *carrog*.

The beast starred in an eighteenth-century burlesque opera, *The Dragon of Wantley*, which was based on a spoof dragon ballad printed in 1699 in *Wit and Mirth; Or, Pills to Purge Melancholy*. At the climax, a Yorkshire dragon dies after a kick by 'a furious knight', More of More Hall, who had consumed a quart of aqua vitae and six pots of ale to summon up sufficient courage to wage battle:

'Murder, murder!' the dragon cried.
'Alack, alack for grief!
Had you but missed that place, you could
Have done me no mischief.'
Then his head he shaked and trembled and quaked,
And down he laid and cried;
First on one knee, then on back tumbled he,
So groaned, kicked, shat and died.

The prestigious scientific journal *Nature* has even carried an article on the ecology of dragons written by the mathematical biologist Lord May, who went on to be the science adviser to the British Prime Minister and is now the President of the Royal Society, London. May observes that dragons are 'both omnivorous and voracious', with great variations in diet: one made do with two sheep every day while another, kept by Pope St Sylvester, consumed 6,000 people daily. Their life span seems to range between 1,000 and 10,000 years. And the reason we don't see dragons today? Driven to extinction by man, of course, with a primary factor being 'commercial over-exploitation, primarily for pharmacological purposes'.

May then attempts to work out where dragons fit into the tree of life and makes the basic observation that the griffin and canonical dragon are six-limbed (four legs, two wings) whereas the wyvern (a heraldic beast) and cockatrice (part snake, part cock) are four-limbed (two legs, two wings). The latter four-limbed arrangement is typical of creatures with backbones (vertebrates) and thus 'wyvern and cockatrice can be envisaged as radiations from the basic vertebrate theme. But dragons, griffins, centaurs and angels belong to an entirely different lineage, the evolutionary history of which is shrouded in mystery.'

We now know why dragons seem so real to Muggles. Although they are ultimately the products of the human imagination, their inspiration came from extraordinary creatures that once lived on earth. Inventing dragons, trolls, giants and other monsters helped our ancestors make sense of some very puzzling observations and discoveries, from horns and giant bones to lights and commotion in the sky, even stones of an unnaturally regular shape. As J.K. Rowling's alter ego Hermione Granger once suggested, legends have a basis in fact. Newt Scamander also observed that various exotic creatures thought to be imaginary now were known to be real in the Middle Ages.

Take the troll as one example. One could speculate that

this creature may have come about when modern-type humans, the Cro-Magnons, arrived in Europe from Africa 40,000 years ago and encountered Neanderthal man. Our Neanderthal cousin became extinct about 28,000 years ago and there is endless debate over the reasons why. Perhaps they live on today in fables. Neanderthals were sufficiently tough and resourceful to be feared, making them ideal material for folklore. The memory of powerful men with beetle brows and big brains who made tools and buried their dead may still linger, thanks to a chain of storytelling that stretches from today to Scandinavian folklore (*troll* is from Old Norse for 'demon') and then on back to those first encounters tens of millennia ago.

Werewolves, Professor R.J. Lupin and other hybrids

Trolls are far from being the only example of how fact becomes fable. Many of the monsters that Harry encounters have inhabited the human mind for millennia. Take Professor Remus Lupin, the werewolf, for example. Archaeologists and anthropologists now believe that our awareness of half-men/half-beasts dates back to the dawn of humanity and the first fully modern people, around 50,000 years ago in Africa.

Evidence of this ancient fascination with monsters emerged from an analysis of ancient rock art sites in Europe, Africa and Australia. The art has one common feature: animal–human hybrids, according to Christopher Chippindale of Cambridge University, who conducted the study with Paul Tacon of the Australian Museum in Sydney.

The surveys included some of the most significant pre-historic art sites, from rock art painted on cliffs in northern Australia to that on ledges in South Africa and inside caverns in France and Spain. Creatures that were hunted locally were often depicted. For example, those at the Grotte Chauvet near the Ardèche gorge in France are more than 30,000 years old and show charging rhinoceroses and horses rearing on their hind legs. But Tacon and Chippindale found only one common element among the 5,000 examples of Stone Age art from the various continents: 1 per cent of images were of 'therianthropes', 'zoomorphs' or 'anthropozoomorphs' – human–animal hybrids, or composite creatures.

The depictions sound like something out of the pages of Harry Potter's copy of *Transfiguration Today*. A feline-headed statuette thought to be roughly 32,000 years old was found in Germany. A bison-headed human and a bizarre mix of human, reindeer, horse and feline features gaze out from the rock face in Trois Frères, France. In Australia paintings in red, orange and yellow mineral pigments depict feathered humans with birdlike heads and 'cone-heads' that could be snake or lizard-headed beings. There are drawings of men with the heads of fruit bats or kangaroos, and bark art that featured mermaids. Because human–animal hybrids are the subjects of the oldest surviving figurative art, the researchers speculate that as soon as archaic humans became fully modern they began imagining hybrids and recording them.

By entering the mindset of the ancients, who communed closely with animals and depended on them for their survival, one can begin to see how such beliefs emerged. Credence in man-monsters is 'at the core of the human psyche' and part of what makes us human, say Tacon and Chippindale. These chimeras highlight our close links with the animal world, which we evolved with and depend on for food and warmth. They hint at a Darwinian understanding of our evolutionary ancestry. They also highlight human frailties and strengths. The man who was portrayed as a bear is broadly equivalent to a superman with great physical strength. Other hybrids reflected our desire to fly like a bird or gallop as fast as a horse.

Half-and-half creatures such as cat-headed women may

also represent key transitions in Stone Age life, commemorating milestones in an individual's social status – for instance, from childhood to adolescence, or from girl to wife, or boy to husband. They may mark seismic shifts in a culture, perhaps the arrival of newcomers or the advent of powerful technology. They may even record changes in the environment. In Australia, for example, a rise in sea level at the end of the last ice age was matched by an increase in rock art.

Most profound of all, these hybrids may illustrate a transition between the real world and other worlds. They express a feeling that there is more to life than material existence. The other worlds are numerous, from heaven and the underworld to those places inhabited by spirits, gods, mythical beings and what Chippindale calls 'ancestor spirits that are mysterious, powerful and dangerous'. Presumably, these are the same alternative dimensions that people can tap into through the use of drugs, rituals or trances.

Neuroscientists would add that the hybrid beings reveal a little about how the brain organises information about creatures. This is highlighted by the case of a man called Philip, who suffered a severe head injury in a car crash that damaged the temporal lobe of his brain. The former trainee draughtsman suffers visual agnosia which leaves him unable to recognise faces – even those of his partner or their daughter. He will look at a sheep, a deer or even a bear and say that it is a fantastical creature, says Roz McCarthy, a Cambridge University neuropsychologist who has studied his case. It is as if Philip has been subjected to a spell that made him forget. Yet he thinks that a montage of rabbit and elephant, a 'bunnyphant', is a real creature. While the depictions of strange hybrids, such as chimeras, centaurs and dragons, are probably not the product of people with agnosia, they do reflect how the brain stores knowledge of natural shapes, forms and structures.

A brief history of chimeras

Once hybrids became established in human consciousness, they remained firmly lodged there. There were the human–animal hybrid gods worshipped by the Egyptians – such as Anubis, the god with a jackal's head, and bird-headed Re-Horakhty, Thoth and Horus. There were satyrs (human–goat) and the centaurs (human–horse) that still roam the Forbidden Forest near Hogwarts. Then there was the Minotaur, a bull-headed man with a taste for Athenian flesh. And, of course, there was the sphinx, which had the body of a lion and the head of a woman.

Later came legends such as the werewolf (before Professor Lupin's condition could be controlled with Wolfsbane Potion, or the Homorphus Charm), and finally specific creations such as Bram Stoker's Dracula, an 'undead' human with batlike features who preyed on the living. Horror films that followed likewise tapped a very ancient urge. Human–animal hybrids have a long history of thrilling and chilling. It is telling that one of the Stone Age animal-head beings studied by Tacon and Chippindale is depicted attacking a woman, reminiscent of a poster for early Hollywood horror films.

The impact made on anyone who saw a representation of a hybrid during a cave visit must have been similar. 'For anyone armed with only a guttering candle, the experience would have been utterly terrifying in the Stone Age. You would crouch down a corridor and would then be suddenly confronted by a half-man, half-lion, or something similar,' says Chippindale. The experience would have been particularly potent when viewed by people in an altered state of consciousness, whether from starvation, drugged with natural hallucinogens, or after dancing themselves into a hypnotic trance.

Modern culture is unusual because it is so materialistic, remarks Chippindale. To think of animals as physical objects, while dismissing these chimeras as imaginary, is not the way that ancestral cultures would have viewed them. In ancient societies there were three classes of being: humans, animals and the 'others'. And it is because of the latter that we still shudder at the thought of walking in the woods at night and Harry Potter fears the Forbidden Forest. Something may lurk there. Indeed, the word 'savage' comes from the Latin *silvaticus*, 'of the woods'.

There are many examples of mixed-up creatures of the kind found in myth and legend that prowl around Harry's enchanted world: Professor Minerva McGonagall, deputy headmistress at Hogwarts, who can turn into a tabby cat with spectacle markings around her eyes; Peter Pettigrew aka Scabbers the rat, who is an Animagus; and, of course, there is the Boggart, which likes to lurk in enclosed spaces. That creature is a shape-shifter with a difference: when confronted, it becomes a physical manifestation of our worst fears.

Griffins and dragons

In classical times, other monsters emerged as people attempted to rationalise the strange, peculiar and downright odd. Not only were Greeks and Romans familiar with the fossils of extinct creatures but they also struggled to come to terms with these puzzling remains of living things, according to studies by the independent researcher Adrienne Mayor. Indeed, they struggled so mightily that many of the mythical creatures that Harry Potter is familiar with could in part be the result of the efforts of ancient people to assimilate the existence of fossil bones scattered throughout sedimentary deposits.

The story of the griffin, a legendary race of four-footed birds with the beaks and wings of eagles and claws of lions, marks one of the best examples of a palaeontological discovery that influenced classical art and literature, and that influence can still be felt today. The creature even appears in the name of Harry's house for example: Gryffindor means 'golden griffin'

(griffin d'or), which harks back to the creatures' ancient association with gold. And Buckbeak, the Hippogriff who escaped with Sirius Black, is a medieval version of the classical griffin, adds Mayor.

The griffin legend originates with Scythian nomads who, in about 675BC, met the Greek traveller Aristeas near the foot of the Altai mountains and told him of a wilderness to the east, where hoards of gold were guarded by the fierce creatures. Aristeas worked these griffins (from *grups*, Greek for 'hooked') into a story in which men on horseback battled the creatures for possession of the gold fields. In later accounts, the griffins protected their nests and young from the goldminers. It is telling, says Mayor, that contemporary accounts of the griffin did not refer to it as the offspring of gods that lived in the mythic past but spoke as though it lived in the present. The griffin soon began to be a routine image in classical art and architecture, and can still be seen today. There is even a brass knocker shaped like one on the entrance to Dumbledore's dwelling.

In her hunt for the griffin, Mayor found a trail of clues that stretches east from the Black Sea to the gold fields of the Hindu Kush, Altai mountains and Gobi Desert. In the shifting sands of the desert, the archetypal griffin is revealed by Mayor to be none other than the dinosaur *Protoceratops*. The entire legend of the griffin can be traced to the creature's remains that litter the desert, according to her book, *The First Fossil Hunters.*

Protoceratops, an eight-foot herbivore, is the most common dinosaur fossil found in the Gobi. Its fossils are easy to spot; there is a distinctive contrast between the bleached white remains and the desert's red sediments. Many aspects of the fossils dovetail with the legend: the beast's beak; its size; evidence of nests; the association with gold deposits (through Scythian goldminers); the way that its frills can break off to leave what looks like ears; and its elongated shoulder blade, which could have been interpreted as an anchor for wings.

Long ago, people dreaded the Gobi. Its sands were littered with the dead, from the carcasses of horses to the remains of their riders. In the thirteenth century, the Chinese had feared the 'fields of white bones'. In the wind-scoured dunes, the potential culprits who caused so much carnage emerged when fossil remains were laid bare by the forces of erosion. For anyone passing by the remains of *Protoceratops*, what more evidence did they need that the desert had a frightful guardian? In this way the legend of the griffin was born.

Given how well established the legend became – even though relatively few people actually saw the remains – what other myths or legends may have been based on fossils? Plenty, says Mayor. Greek myths are full of heroes and giants, reflecting how people in classical times were constantly confronted by the remarkable petrified remains of the giant creatures that once roamed the earth.

The anthropologist Stewart Guthrie of Fordham University has argued that there is a powerful instinct to anthropomorphise. When vast bones were ploughed up by Roman farmers, which often happened, according to the poet Virgil, they were usually thought to be the remains of ancient warriors. The fossilised skeletons of giant giraffes, mastodons, mammoths and other extinct behemoths captured the imagination of ancient Romans and Greeks eager to find evidence of primeval giants and monsters. They treasured these perplexing bones and put them on display in temples and other public places. The emperor Augustus (63BC–AD14) even set up the world's first palaeontological museum at his villa on Capri; his biographer, Suetonius, reported that it housed giants' bones and the weapons of ancient heroes.

The ancients would not have been surprised to encounter someone the size of Hagrid. While men at that time stood about five feet tall, the ancient heroes seemed to be three times taller, and it is no coincidence that a mammoth femur is three times the size of a human thigh bone. Pliny the Elder would remark that, from the evidence of such bones, 'the stature of the entire human race is getting smaller'.

Mayor cites many other impressive examples to back her theory. A fossil elephant shoulder blade may have been enshrined as the shoulder bone of the giant hero Pelops in antiquity. Remains that the Lydians thought were the ogre Geryon and his outsize cattle were probably extinct mastodons and bovines whose fossils lie scattered around eastern Turkey. In Plymouth Hoe, Britain, when gigantic jaws and teeth were dug up they were thought to belong to the legendary giant Gogmagog.

Attributing marks in rock to the steps of fabulous ancestors was also common. From classical times, the tracks of prehistoric creatures have been billed as the footprints of gods, giants, fairies, angels and devils and the thoroughfares of heroes. Herodotus, in about 450BC, was shown a three-foot-long footprint supposedly left by Hercules, for example. Other tracks claimed to be those of the hero and found in the heel of Italy are probably the ancient footprints of a cave bear, which can be readily mistaken for those of humans, says Mayor.

Dragons may also have had prehistoric origins. The palaeontologist Kenneth Oakley suggested that features of the Chinese dragon mimicked the features of prehistoric mammals of China and Mongolia, such as distinctive antlers resembling those of fossil deer. The Siwalik fossil beds, which lie in foothills that stretch from Kashmir to Nepal, could have inspired the sightings of dragons of huge size and variety that Apollonius of Tyana claims prowled around northern India. And dinosaur tracks across the red Triassic sandstones in the Rhine valley could have laid the foundations of the legend of the slaying of the dragon Fafnir by the hero Siegfried. Indeed, the English word *dragon* is a translation of a Latin term used in the Middle Ages to describe the fire-breathing, flying reptilian monster of Germanic myth.

Meanwhile, in medieval Europe, the huge bones of prehistoric animals were thought to belong to saints. Mammoth and rhinoceros skeletons were propped upright and claimed to be giants, primitive cavemen and even barbarian warriors such as Visigoths. In northwest Australia, the Aborigines believed that the three-toed tracks of big flesh-eating dinosaurs were left by Marella, a giant 'emu man'. In Malta, grooved five-point impressions said to be the devil's footsteps were actually fossil traces of sea urchins.

The basilisk

Dragons are not the only creatures from the Harry Potter books who could have been inspired by fossils. Another could be the basilisk, the king of serpents (from *basileus*, king), which featured in classical legend and could kill with a whiff of its toxic breath or a poisonous glance. In J.K. Rowling's description, the serpent was poisonous green and as thick as an oak trunk, with a blunt head covered with scales. Harry encountered it in the Chamber of Secrets.

A similar supernatural creature can be found throughout aboriginal Australia, with some depictions dating back 6,000 years. Called the Rainbow Serpent, this ancestral being sometimes had a kangaroo head and was responsible for great acts of both creation and destruction. The serpent lived at a period of ancestral prehistory referred to as *garrewakwani*, The Dreaming, which saw half-men, half-animal (bird or fish) that paved the way towards the evolution of humans.

Evidence of an awesome antecedent of the basilisk was found in part of northern Africa now called the Faiyum Depression of Egypt, a low-lying area between the Nile Valley and the Gebel el Qatrani escarpment. Though it is desert today, it was once lush and tropically vegetated, with lake, river and swamp. Fossil remains of the prehistoric fen and its exotic inhabitants lie buried beneath the sand.

Palaeontological discoveries at the site include hippopotamus-like and rhino-like creatures; hyena-like hunters and scavengers; unusual elephant-like beasts; ancestral whales that still had limbs; turtles; sea cows; birds and rodents. Among them lurked a relative of the basilisk. Called *Gigantophis* (giant snake), it is a constrictor-like creature that was possibly the largest snake, which slithered around what is now Egypt and Libya between 35 million and 45 million years ago.

The name 'Gigantophis' is appropriate as any single vertebra of the beast is bigger than the largest known anaconda vertebra. 'I suspect that a large *Gigantophis* would be about 35 feet long,' comments snake expert Jason Head of Southern Methodist University, Dallas, Texas. *Gigantophis* was related to the modern boids, or boas and pythons, which still have tiny clawlike toes that are remnants of ancestral legs. The great serpent probably also had these 'spurs', which are actually minute femora (thigh bones). 'Boas and pythons use them in mating, where the male will rub the female with the spurs to initiate copulation,' says Head.

We know, of course, from Harry Potter that one awesome *Gigantophis* remained hidden in the Chamber of Secrets under Hogwarts. We also know that the basilisk had a venomous bite and that, if it had not been for the tears of Fawkes the phoenix, Harry would have perished. However, Head says that, given the types of snake that have venom today, he doubts that *Gigantophis* had a poisonous bite.

J.K. Rowling mentions the murderous stare of the basilisk. Perhaps this refers to heat-sensitive pits on the front of the snake's face, which it used for finding its warm-blooded smaller prey at night. '*Gigantophis* was, as the name indicates, a huge snake, and likely ate large-bodied prey (but I'm at a loss to tell you specifically what prey),' explains Head. 'As a result, it probably hunted primarily by chemoreception [scent], because large prey are more easily tracked by scent than small prey, and can be tracked from a greater distance from scent than heat.'

Comets, dragons and King Arthur

Not all physical evidence that inspired the ancients to believe in dragons and giants came from living things, however. Some say that Cornwall has more stories about giants than any other English county because local people were in awe of its great number of megalithic monuments. In both Norse and English traditions, edifices of stone were said to be 'the work of giants'.

Cosmic phenomena also provided inspiration. Folklore suggests that the death of King Arthur – in either 539 or 542 depending on your source – plunged Britain into a dark age. By studying tree rings, and looking specifically for thinner rings which are linked with meaner climates, Mike Baillie at Queen's University Belfast has found that an environmental downturn struck about the same time, lasting from 536 until 545. Looking in Arthurian legend for clues to the cause of the disaster, Baillie suggested that earth may have had a close brush with a comet that loaded the atmosphere with dust and debris, bringing gloom and subsequent crop failure. No wonder then that Merlin, Arthur's magician, is depicted in mythology as a 'red fiery whooshing dragon flying in the sky' who throws thunderbolts.

There are other examples where ancient myth may have been inspired by the visit of a comet, raining fireballs and meteors. The Book of Revelation mentions a huge, red, multiple-headed dragon in the sky. Babylonian mythology refers to *tiamat*, representing primeval chaos, which gives birth to giant snakes, a horned serpent and dragons.

As well as its destructive tendencies, nature's constructive tendencies also led to legends. Celts believed that the Giant's Causeway in Northern Ireland, an amazing structure consisting of around 40,000 geometrically shaped basalt columns, was the work of one of Hagrid's relatives. The giant Finn McCool supposedly built the causeway to reach the hideaway of a Scottish giant, Benandonner, who lived on the island of Staffa.

The first hint that the giant Ulster warrior was a fabrication came in 1693, when an expedition report to the Philosophical Society of London concluded that the mysterious formation of basalt columns was in fact the result of natural causes, though it was not known precisely which ones were involved. Even recently, the columns were thought to have been formed by cooling lava, but the actual process that produced their very regular polygonal shapes has never been completely understood – until now.

The answer may at last have been provided by Alberto Rojo of the University of Michigan and Eduardo Jagla of the Centro Atomico Bariloche in Argentina. In their version, the causeway formed when, as the lava cooled, fractures formed at the surface and moved downward. At the top, the fracture pattern looked like the familiar arrangement that we see in drying mud or in paint – a random pattern of curved lines. But as the fractures moved downward they began to take the path of least resistance. It turns out that the configuration that minimises the energy required to penetrate the interior is a regular pattern of hexagons, pentagons and heptagons. These deeper regular patterns – the Giant's Causeway – have since been laid bare by erosion.

The researchers support their theory with a computer model of the process backing their mathematical model, along with a simple experiment with cornstarch that anyone can do at home. Although the frontiers of modern physics are thought to lie in the quantum realm of the very small, and the relativistic realm of the very big, Rojo believes that fundamental open questions and mysteries still remain at more familiar dimensions. Even without Finn McCool, there remains a great deal of magic in everyday experience.

Hunt for the werewolves

The menagerie of mythical beasts has had another source of inspiration in nature to fire the fertile human imagination: misfortune. Over the ages, disorders, genetic defects and disease may have created other frightening beings that came to populate legend and the human psyche.

The legend of the werewolf, for example, may be linked to a psychiatric disorder known as lycanthropy. There are many documented cases of people deciding that they are wolves or other ferocious wild animals, although whether the legend or the personality disorder came first is unclear.

Others have attributed the werewolf legend to a rare disease called congenital generalised hypertrichosis. Sufferers have excessive amounts of hair on their face and upper body. The genetic mutation is of keen interest because it could shed light both on how to grow hair and on human origins. Presumed to be an atavistic genetic defect, it has been speculated that it reactivates a gene that has been suppressed during evolution.

Atavistic defects would have been as fascinating to an ancient mind as to a modern geneticist. They include the third nipple seen on some people, an extra digit on the feet of horses (whose ancestors had more toes), the occasional appearance of hind limbs on modern whales (even though their ancestors moved from land to water between 40 and 50 million years ago), and maybe even body hair from head to foot of the kind once worn by our ancestors. Verified victims of congenital generalised hypertrichosis have often worked in circuses as 'ape men' or 'werewolves'. Other sufferers could have been the hirsute Weird Sisters, who played at the Hogwarts Yule Ball.

The hypertrichosis disorder was formally identified in a Mexican family and named in 1984 by José María Cantú's team at the University of Guadalajara. A gene hunt was launched by an international team led by Pragna Patel of the Baylor College of Medicine in Houston. At the time of writing, the werewolf gene still eluded her. A successful outcome would not only shed light on hypertrichosis but could also explain the ability of Harry Potter's hair to grow back rapidly.

The influence of real world genetics on mythical creatures does not end there. Vampires and werewolves may also be linked by a recent discovery that, no doubt, is discussed in Gilderoy Lockhart's *Voyages with Vampires*. A suggestion for their medical origins was proposed in 1985 at a meeting of the American Association for the Advancement of Science by David Dolphin, a biochemist at the University of British Columbia. Perhaps these individuals had porphyria, a genetic disease that is also thought to have struck King George III.

This malfunction in the body's chemistry makes the skin sensitive to light, which explains why sufferers tend to avoid the sun. The chief defence against the painful effects of the sun, Dolphin suggests, would be to venture forth only at night, as werewolves and vampires were said to do. Some victims of the disease also become very hairy, he says, conceivably one of nature's efforts to protect the skin from the sun.

The disease can also cause retraction of the gums, making them so taut that the teeth, although no larger than ordinary, jut out in a menacing manner akin to something more animal than human – these teeth further nourishing the vampire and werewolf myths. And the disease may explain why vampires, or porphyria victims, might well have been afraid of garlic, in accord with mythology. Garlic, Dolphin says, contains a chemical that exacerbates the symptoms of porphyria.

Dolphin also notes that one treatment for the disease is an injection of a blood product, heme. Since that treatment did not exist in the Middle Ages, when the myths originated, he speculates that victims might have instinctively sought heme by drinking blood, as was supposedly the custom of vampires. Voldemort, Harry's nemesis, had a taste for unicorn blood and it is interesting to speculate whether he, too, was a sufferer.

This is by no means the end of scientific musings on the

reality behind vampires. In 1998, Spanish neurologist Juan Gomez-Alonso linked the legend to rabies. The symptoms include insomnia, an aversion to mirrors and strong smells (though not specifically to garlic), and an increased sex drive. And rabies, of course, is transmitted by biting, reflecting how the victims of vampires usually become vampires themselves.

Zombies

I had almost lost the sense of feeling nor could I distinguish between light and heavy objects, a quart pot full of water and a feather was the same in my hand. We each took a vomit and after that a sweat which gave great relief.

Captain James Cook, on suffering fugu poisoning, 1774

Zombies are supposed to be the 'living dead', and sound a little like the dreaded Dementors who guard Azkaban. They do make the occasional appearance in the Potter books. For chasing away one particularly troublesome zombie, Quirrell was supposedly given a turban by an African prince.

It is surprising that Hogwarts has not been bothered more by zombies. By one account, around 1,000 cases of zombification are reported each year in Haiti. Many explanations have been put forward for the condition. Doctors blame poisoning. The clergy believe that its victims are genuine targets of voodoo sorcery. Others believe that mental illness is to blame. The details are bizarre and contain echoes of how Voldemort's spirit was separated from his body when Harry's mother sacrificed herself to save her son and deflected the Dark Lord's curse.

In a typical zombie scenario, a young person suddenly falls ill, apparently dies, and is placed in a tomb. A sorcerer (*boko*) steals his remains, and the cadaver is secretly returned to life. However, there is a catch. Local beliefs separate the physical body from the animating principle and agency, and from awareness and memory (*ti-bon anj*). The latter is removed and stored by the sorcerer, where it is known as the *zombi astral*. What is left is an animated body without will, a zombie who will do the bidding of the *boko*.

One theory for zombification, put forward by Wade Davis, is that the *boko* uses a neurotoxin. Davis studied the strange case of a man claiming to be Clairvius Narcisse, a Haitian man who was pronounced dead by physicians in 1962. According to Narcisse's account, he was merely paralysed. He claimed he heard himself declared dead, and remembered being placed in a coffin and buried. He was later dug up, beaten and forced into slavery under a voodoo 'zombie master'.

Davis found that the *boko* used a powder made of ingredients that included porcupine and puffer fish, both of which are known to contain tetrodotoxin. This potent marine neurotoxin is named after the order of fish from which it is most commonly associated, the Tetraodontiformes (*tetras*: four, and *odontos*: tooth). The toxin is not actually produced by the fish themselves but by symbiotic bacteria that live within them: as a consequence, puffer fish grown in culture do not produce tetrodotoxin until they are fed tissues that contain the organism from a toxin-producing fish.

In Japan, puffer fish are a delicacy. Trace amounts of tetrodoxin in the fish – *fugu* – are thought to induce a tingling of the extremities, warmth and euphoria. But the fish must be prepared by licensed chefs with the knowledge and ability to remove all the toxin-containing parts, notably the liver and ovaries.

The toxin blocks channels in the membranes of neurons (nerve cells) that allow charged sodium atoms to pass through, upsetting the delicate chemical balance required for them to work properly. When the channels open there is a nerve impulse, a change in voltage across the nerve cell wall. The impulse travels along the length of a nerve because the channels open one after the other in a kind of domino effect. But when blocked by tetrodotoxin, the nerves cannot function.

The first sign of puffer fish poisoning is a slight numbness of the lips and tongue, appearing between twenty minutes and three hours after eating a tainted morsel. Sensations of lightness or floating follow, along with headache, nausea and vomiting, until paralysis begins to set in. The victim, although completely paralysed, may be conscious and in some cases completely lucid until shortly before death. Severe cases look blue, and the eyes become fixed and dilated. Vital signs may cease to be apparent even to an experienced physician. In other words, the victim *seems* to be dead. There are also suggestions that the victim is then revived and controlled using zombie cucumber – *Datura stramonium* – though the plant is listed as containing atropine and scopolamine, alkaloids that in themselves can cause delirium, delusions, hallucinations, disorientation, incoherent speech, even seizures, coma and death.

Other insights emerged from a study of three cases of zombification by Roland Littlewood of University College London and Chavannes Douyon of the Polyclinique in Port-au-Prince, Haiti, which was reported in the *Lancet*. They found evidence that the breaking open of tombs by *bokos* is widespread and noted that the majority of temples they visited contained human skulls and other body parts, suggesting a culture of body snatching. However, when they conducted DNA tests on two of the victims they found that one, WD, was not the son of his putative parents, nor was the other, MM, related to either of the men who claimed to be her brothers.

They concluded that one ingredient of the strange phenomenon is the way that people with schizophrenia, brain damage or learning disabilities seen wandering in Haiti are mistaken for zombies. 'They are not feared in Haiti at all. The fear is of the *boko* and of being zombified oneself,' explains Littlewood. 'Soul gone, nor properly dead, it's very like the European fear of becoming an earth-bound spirit.'

The second ingredient was a willingness of families to take the victims in. 'Mistaken identification of a wandering, mentally ill stranger by bereaved relatives is the most likely explanation – as in the cases of MM and WD.' However, they were puzzled why a young person adopted this way by a family would agree not only to be a zombie but also agree to become a relative.

They concluded that there is probably no single explanation for the phenomenon: 'Given that death is locally recognised without access to medical certification, and that burial usually occurs within a day of death, it is not implausible for the retrieved person to be alive. The use of *Datura stramonium* to revive them, and its possible repeated administration during the period of zombie slavery, could produce a state of extreme psychological passivity.'

Fairies, pixies and elves

Little people, such as fairies, pixies and elves, may also exist. They are certainly common in legend. In Cornwall, a race of fairies called spriggans guarded treasure. Others stole food when invisible. Francis Grose wrote in his *Provincial Glossary* (1787) that they also like making cakes 'in the doing of which they are said to be very noisy'. Somewhere off Milford Haven lie the Green Islands of the Sea, said to be the invisible country of the fairies.

The last home of the fairies in England was supposedly Harrow Hill, Sussex. However, some may live on, in the form of sufferers of a genetic disorder called Williams syndrome, according to Howard Lenhoff of the University of California, Irvine, whose daughter, Gloria, has the condition. First described by the eponymous heart specialist in New Zealand in 1961, Williams syndrome affects around one in 20,000 births.

It is caused by the loss of twenty specific genes lying on one of the pair of chromosome 7 in cells. Those with the disorder have cardiovascular problems such as heart murmurs and sub-normal intelligence, and are acutely sensitive to sound. They are loving, caring and sensitive to the feelings of others. Despite having low IQs, many are good storytellers and have a talent for music, notably perfect pitch. Most striking of all is their appearance. A relatively large number are short. They have child-like faces, with small, upturned noses, oval ears, and broad mouths with full lips and a small chin. They look and behave like the traditional depiction of elves. 'The "wee people" of folktales often are musicians and storytellers,' says Lenhoff. 'Fairies are frequently referred to as the "good people" and are said to repeat the songs they have heard, and can enchant humans with their melodies. Much the same can be said of people with Williams syndrome.'

The Muggles' magical menagerie

Hogwarts does not have a monopoly on strange creatures. Dragons, giants, zombies, werewolves, pixies, elves and witches also inhabit the Muggle world, and all have one thing in common: they have been put forward as explanations for natural events, whether the bleached bones of long-extinct animals, genetic misfortune or strange rock formations. Given what we know about the fertile human imagination, it is perhaps more surprising that we have not inherited an even bigger magical bestiary from our ancestors.

In this exploration of the history of magic, the natural world has also been the source of a different kind of inspiration for witchcraft and wizardry, for it has provided a vast range of plants that have the power to harm and to heal. In traditional magic, they were used alongside formulations that, to Muggle scientists at least, were primarily symbolic: a magician could try to hurt someone by destroying a sample of his hair; a person might swallow gold to cure jaundice, or place a stone in a tree fork to delay the setting of the sun. Unlike these superstitions, however, Snape's plant-based potions exert a power that can still be felt today, as drug companies race to uncover their secrets before many plants and flowers vanish.

CHAPTER 11

The Potions Master

Double, double toil and trouble;
Fire burn and cauldron bubble.
Fillet of a fenny snake,
In the cauldron boil and bake.

Shakespeare, *Macbeth*

Remarkable brews made with bizarre recipes have always been a part of the magician's arsenal. During Professor Severus Snape's potions classes in the dungeons of Hogwarts, Harry Potter learned much about herbs, fungi, puffer fish eyes and other peculiar ingredients. Surrounded by wooden desks, steaming cauldrons, brass scales and jars, Harry probably tried his hand at making Swelling Solution and Sleeping Potion, among many others, and pored over standard textbooks such as *Magical Drafts and Potions*. No doubt many of these concoctions were also brewed up by mediwizards and Madam Pomfrey, the Hogwarts matron.

Ancient books written by Muggles reveal much about the kinds of plants, herbs and other ingredients traditionally used by witches and wizards. The recipe of perhaps the best-known witches' brew of all is that featured in Shakespeare's *Macbeth*. As well as the fillet of fenny snake there are toxic ingredients, such as root of hemlock and slips of yew, along with exotica such as scale of dragon and witches' mummy, the latter no doubt reflecting the ancient belief that the desirable qualities of some object, such as the longevity of a mummified body, might in some way be absorbed by eating it.

How did Snape's ancestors and those who brewed potions in the Muggle world come by this knowledge? That wizard authority Kennilworthy Whisp describes the visit of a team of European herbologists to New Zealand in the seventeenth century to research magical plants and fungi, but the origins in fact date back much further than this. Amazingly, this particular form of witchcraft was probably born before modern man even walked on the planet. That surprising fact may lead a sceptical reader to question whether these potions work at all. While Muggles have found that, thanks to the conjuring abilities of the human brain, it does not always matter, not all ancient poultices, brews and serums relied on mental wizardry alone. Some did have very real and very powerful effects on an ailing body. Their legacy lives on both in St Mungo's Hospital for Magical Maladies and Injuries and the global pharmaceutical industry. Both exploit nature's storehouse of medicines, but perhaps not for much longer. The planet's burgeoning Muggle population is wiping out rainforests and other complex habitats before we have had a chance to study the plants, insects and microbes that live there. Some of the most powerful potions of all may never be discovered.

How drugs work

Muggles understand the action of drugs and potions through the efforts of witches, wizards and other experts who study chemistry, molecular biology and pharmacology. Although the word 'chemical' is often used as a pejorative catch-all for toxins and carcinogens, it also describes any and every life-saving drug and every substance in your body, whether they are coursing through your veins, sticking cells together or providing a window for the eye, or to the soul, for that matter.

Any medicine, whether an ancient poultice or a modern gelcap, contains chemicals, collections of atoms bound together to form molecules. They owe their medicinal properties to the way they interact with other molecules in the body: receptors; proteins in the membranes of cells that usually respond to hormones; enzymes, another type of large protein molecule that speeds chemical reactions within cells; pores in the walls of cells called channels; or even the genetic materials of DNA and RNA which carry instructions to make the protein building blocks of cells.

The reason that Snape used plants for many of his brews and poultices comes as an indirect result of their being relatively immobile. Time-lapse films reveal how the imperceptible movements of plants enable roots to snake through soil, flowers to furl and unfurl, and plants to swivel to face the sun. But they can't really wriggle, walk or fly about like worms, humans and birds (except, perhaps, for the Whomping Willow).

Lacking locomotion, flowers have developed attractive scents and fruits that offer sweet nourishment to establish contracts with other species with the means of transport to spread their pollen and seeds. Because they can't run or hide, plants have also developed a fearsome range of chemical weapons to deter, maim and even kill insects and animals that eat them. We have learned to exploit this impressive range of chemicals. Ethnobotanists Michael Balick and Paul Alan Cox estimate that we depend on the plant kingdom for about one quarter of prescription drugs and almost all our recreational ones, 'including the caffeine in coffee, the nicotine in tobacco, the theophylline in tea, the theobromine in chocolate, and a virtual cornucopia of other psychoactive substances'. These include, of course, some of the hallucinogens we encountered earlier.

For obvious reasons, exploitation of the medicinal properties of plants dates back long before Hogwarts was established. They would have boosted the survival of our ancestors but so too would plant chemical defences such as nicotine and cocaine, argue Roger Sullivan of the University of Auckland and Edward Hagen of the University of California, Santa Barbara. At first plants evolved these defences to mimic chemical messengers, called neurotransmitters, and so disrupt signalling processes between nerves. But our ancestors evolved detox machinery in the liver – notably an enzyme called cytochrome P-450 – and a way to exploit these neurotransmitter substitutes. In short, Sullivan and Hagen believe that taking stimulants and narcotics could be a part of our evolutionary inheritance that dates back millions of years.

Stimulant alkaloids, such as those taken in by chewing coca, tobacco and khat leaves, or indeed the betel nut (the fourth most commonly used drug in the world), could have helped our ancestors to cope with fatigue and endure harsh environments. Australian Aborigines used pituri, a native plant with a high nicotine content, to deaden hunger cravings. In deserts and alpine ranges, plant drugs could have made up shortfalls of neurotransmitters – such as noradrenaline, dopamine, serotonin and acetylchonline – from a poor diet. Betel raises acetylcholine levels; khat boosts noradrenaline, dopamine and serotonin; and the cocaine in coca increases noradrenaline and dopamine. Sullivan points out that some indigenous people even 'freebased' drugs by chewing them with an alkaline substance such as lime or wood ash to release the free form of the drug so it could be readily absorbed into the bloodstream.

For the diet-conscious, plant extracts were also used for hundreds of years to suppress appetite. Some were contraceptives. Control of flatulence and bowel habits was another ancient obsession and senna was once known as the 'guardian of the royal bowel movement'. The euphoric and painkilling properties of the opium poppy were probably known to the Sumerians as early as six millennia ago. Remedies were also available for the treatment of boils, burns, and other wounds liable to become infected. But there was, of course, also a dark side to Snape's potions, for while some could heal, others would harm and at times kill.

The enchantment of nothing

Any one of the herbs used in magic and witchcraft could have tapped a mysterious healing force that, even today, remains a topic of hot debate: the placebo. A placebo (Latin for 'I shall please') is an inert substance or sham therapy, usually given to highlight the effects of real treatment, but also for its own benefits.

Many scientists believe that placebos offer a plausible explanation for such phenomena as faith healing, 'the power of suggestion', the laying on of hands, a medicine man's poultice, even the healing effects of a mother's kiss and soothing murmur of 'there, there, there'. Indeed, even a cry of 'Abracadabra!' could help an ailing patient as long as that patient had good reason to believe that it was truly a powerful word. In some mysterious way belief seems to be able to help boost the body's recuperative powers – healing wounds faster and boosting the immune system's attack on infection.

The effect was originally noted in a 1955 paper, 'The Powerful Placebo', by Henry Beecher, an anaesthesiologist at Massachusetts General Hospital in Boston. He reviewed fifteen published studies and found evidence of that powerful placebo's abilities in treating postsurgical wound pain, headaches, seasickness, coughs, anxiety and other nervous disorders. Since then, the medical literature has reported even wider evidence of the placebo effects. Between 30 and 60 per cent of patients with illnesses ranging from arthritis to depression report a substantial improvement in their symptoms after receiving a placebo. It is not clear that a placebo can 'cure' any illness, but the placebo effect is impressive when it comes to improving symptoms and reducing suffering.

Placebos have helped patients with hypertension, depression, acne, asthma, colds, arthritis, ulcers, headache, constipation and even warts. Indeed, it is one sign of the importance of the placebo effect that no test of a new drug or other treatment is considered truly valid unless it compares the new agent with a placebo. (In this case, the placebo must look like the real treatment to both doctor and patient but have no pharmacological properties.) Only if the treatment does better than the placebo is it taken seriously.

Not everyone believes in placebos, however. The *New England Journal of Medicine* published a sceptical analysis by Asbjørn Hróobjartsson and Peter Gøtzsche of the University of Copenhagen, who scoured the world's publications for well-designed studies that included not just a placebo group but a group that received no treatment whatsoever. They found 114, which involved about 7,500 patients with forty different conditions. When they analysed the data, they concluded that patients given nothing improved just as much as those given placebos. There were exceptions, notably in smaller trials that were more at risk of bias. The team also found a positive effect in some pain studies, though Hróbjartsson argued this was usually when patients were asked questions such as, 'How much pain do you feel?' where the answers were difficult to quantify precisely and, consequently, more easily skewed.

Perhaps the placebo effect is nothing more than an illusion that gives that all-important sense of control back to patients. The idea that the mind can control symptoms and disease is appealing, and some sceptics have billed it as a secular religion which, like other types of faith, is not going to be shaken by evidence that it is illusory, even if that evidence is published in the *New England Journal of Medicine.*

However, doctors who believe in the placebo have a rational riposte to the *NEJM* paper, arguing that the design of the Danish study lumped together too many diverse conditions to be meaningful. They point out that there was a significant placebo effect for headache, chronic pain, arthritis, depression and many other conditions. As one put it, Hróbjartsson and Gøtzsche did not prove that placebos do not improve anything, but rather that placebos do not improve *everything*. Another agreed that there is no case to use a placebo as a treatment in its own right, but it can help boost the effect of conventional treatments.

Proponents of the placebo effect are strong advocates of the power of belief, and so hold that a patient's mind can affect the working of his body. For example, there is a wide range of evidence linking disease to chronic stress, so it is not a great leap to conclude that ways to alleviate stress could be beneficial. 'The placebo effect will only work on patients who feel ill enough to will themselves better,' says Jon Stoessl of the University of British Columbia in Vancouver.

Stoessl has moved the debate about the mysterious placebo effect to a new level by providing hard evidence that the effect is real and physiological. He found that a placebo can produce measurable effects in the brain for the treatment of Parkinson's disease, the neurodegenerative disorder that causes impaired coordination and tremor as it destroys brain cells that produce the messenger chemical dopamine.

Other studies back this finding, such as one of fifty-one patients with major depression conducted at the University of California, Los Angeles, by Andrew Leuchter and colleagues. The team assigned each patient to one of two double-blind placebo-controlled studies, using either fluoxetine (Prozac) or venlafaxine (Effexor) as the active treatment. Overall, 52 per cent (13 of 25) of the subjects receiving antidepressant medication responded to treatment, while 38 per cent (10 of

26) got better with the placebos.

During the trial, the team used a method called quantitative electroencephalography imaging to examine brain electrical activity. Patients who responded to medication showed suppressed activity of a region of the brain called the prefrontal cortex, while those who responded to the placebo showed increased activity in the same region. 'People have known for years that if you give placebos to patients with depression or other illnesses, many of them will get better,' says Leuchter. 'What this study shows, for the first time, is that people who get better on placebo have a change in brain function, just as surely as people who get better on medication. We now know that placebo is, very definitely, an active treatment. If we can identify what some of the mechanisms are that help people get better with placebo, we may be able to make treatments more effective.'

The same holds true for using placebos to treat pain. Martin Ingvar of the Karolinska Institute, Stockholm, and colleagues used a brain scanner method, called positron emission tomography, to study how test subjects reacted to painfully hot and non-painfully warm sensations, after receiving either a real opioid painkilling drug, a placebo substitute or no treatment at all. The subjects got relief from both the drug and the placebo, both of which increased the activity in a brain region called the rostral ACC. Because pain relief from placebo and from opioid drugs seem to involve some common brain sites, the team believes a placebo can activate the brain's opioid system as well, a natural mechanism that evolved to damp down pain after it has warned the body of danger, so that pain does not become a distraction when it comes to reacting to that danger.

The very fact that believing in something with no possible medical benefit may help treat disease could shed light on why witch doctors, healers and shamans traditionally cast spells. Words can calm people, can make them fall in love, can whip them up into a frenzy, can turn them into killers. The ancients probably thought that it was no use smearing a particular greasy gruel on yourself or gulping down a potion unless you muttered the right words, or incantation, at the same time to boost the effect. And, as Stoessl's work shows, the placebo aids those who expect a reward.

The placebo effect may also shed light on why individual herbal remedies are put to so many different uses. Perhaps it does not matter what the remedy is used for, so long as the patient believes it will be effective. In Argentina, aloe vera is traditionally used to induce abortion. Yet in Bolivia, it is used for constipation, in the Canary Islands for diabetes, in India as an aphrodisiac, in Panama for stomach ulcers, in Peru for asthma, in Puerto Rica for colds, in Saudi Arabia for piles, in South Korea as a contraceptive, in Taiwan for hepatitis and in the West Indies to prevent syphilis. The plant may have some real underlying actions, but the mysterious power of the placebo is the easiest way to account for the diverse virtues of this wonder drug.

Snape's hairy ancestors

When it comes to many herbal medicines, there is more to their action than the placebo effect. This much was also clear to our hairy ancestors, who could detect when a plant had a pharmacological action by a process of trial and error. They made advances in understanding the medicinal properties of plants long before Snape mixed his first potion.

Apes tend to be fussy about what they eat but, over the millennia, have been driven by hunger to try new foods and, in the process, discovered health benefits as well as enduring some stomach upsets. As a consequence, apes were using Nature's pharmacy long before the first wizards. When the practice spread to their relatively hairless peers, it planted the seeds of medicine.

In the 1960s, Jane Goodall, the veteran chimpanzee specialist, spotted chimps swallowing whole leaves in Gombe Stream National Park. Since then, around thirty-four different species of plants have been found to be used by the animals in thirteen great ape study sites across Africa. All the plants have one common feature: bristly, rough leaves. In the neighbouring Mahale mountains, leaf-swallowing has been found to reach a peak about four to eight weeks after the rainy season has begun, timing that is significant: it coincided with a peak of infection with a parasitic worm. The leaves are effective at giving relief because they are not chewed by the chimpanzees and, as a result, rapidly pass undigested through the gastrointestinal tract and are purged in diarrhoea. In the dung of sick apes that eat the leaves, live worms can be found among the excreted leaves.

This is a physical, rather than pharmacological, effect. But there is also evidence of the latter, which scientists call called zoopharmacognosy. Michael Huffman of the Primate Research Institute, Kyoto University and Mohamedi Seifu Kalunde, Tanzanian National Park game officer and medicine man, happened to be watching Chausiku, a constipated chimpanzee in the Mahale Mountains of western Tanzania, and saw her reach out for the shoot of a noxious tree that chimps would normally avoid. She peeled it and sucked its bitter pith. Within a day, her constipation was eased, and her appetite and strength were visibly restored.

The observation, which Huffman reported in 1987, marked the first time a scientist had seen a sick chimp select an unsavoury plant known by humans to have medicinal properties and use it to recover from disease. The pith, from the tree *Vernonia amygdalina*, turned out to contain compounds active against many of the parasites responsible for malaria, dysentery and schistosomiasis, as well as being antibacterial. Local people, the WaTongwe, use the same plant – *Mjonso* – to treat the same illnesses, and it takes them about the same time to recover as the apes. Ugandan farmers feed their pigs young branches of the plant to rid them of intestinal parasites. A number of bitter *Vernonia* species found across Africa, the Americas and Asia are known for their effectiveness against stomach bugs, including parasite infections. Suppressing disease using a cocktail of antiparasite drugs, which thwarts the development of resistance, could be a useful route to new medicines for humans and livestock, rather than relying on the single-action approach typical of a modern drug.

Ape self-medication may hold the key to finding the evolutionary origins of human medicine. Plant toxins usually taste bitter, a flavour to which most animals have an aversion, so selecting these potent leaves must be a learned behaviour, just as humans learn to like, indeed even love, bitter tastes such as Campari, beer or coffee. Like any drug, medicinal plants should not be overused. But this presents no problems to great apes who, like us, are smart enough to learn from their peers the symptoms, the best medicine and the dosage necessary for the desired effect. There is also evidence that some even eat clay as an antidote.

Apes may also have been the first to enjoy herbal highs, pre-dating the consumption of butterbeer by a long time. Huffman, working with Don Cousins, author of *The Magnificent Gorilla*, went through studies of food consumed by gorilla groups in Africa to evaluate their possible medicinal contribution of the diet for preventing disease and, to their surprise, came up with a range of examples suggesting that apes may indulge in plants for their recreational value, from a pick-me-up of the kind found in coffee to hallucinogens that can shake their grip on reality.

African apes eat a number of cola (*Cola*) trees (*Sterculiaceae*), particularly their seeds and fruits. In fact, *C. pachycarpa* is known as 'cola of the gorillas'. People throughout tropical West Africa hold the cola nuts in high regard as charms and remedies, as amulets and as aphrodisiacs. White or light-coloured nuts effect 'love magic', for example, while red has the opposite effect.

When apes eat the seeds, researchers believe that they are having the equivalent of a cup of tea or coffee. Among their human peers, the reputation of the nuts for suppressing fatigue and promoting endurance is legendary. The seeds do not have much protein value and are probably prized for the stimulating caffeine and theobromine that they contain.

Cousins and Huffman found that two hallucinogenic plants are ingested by gorillas in Equatorial Guinea and by chimpanzees in the republic of Guinea: *Alchornea floribunda* and *A. cordifolia* (*Euphorbiaceae*). Gabonese cults use *A. floribunda* root to become drunk and amorous. It is said to provide a state of intense excitement followed by a deep, sometimes fatal depression. 'It is a tantalising thought that gorillas might be directly affected by these same properties,' says Huffman.

The apes even resort to a root that is being studied by doctors for use as a drug detox. The *Tabernanthe iboga* root has been exploited by gorillas of Sindara on the Ngounie river, south of Lambarene, Gabon. In southeastern Cameroon gorillas eat the flowers and branches of iboga, while in Gabon they eat the fruits, stem and root of the plant. *Tabernanthe manii*, a related species which probably has a similar makeup, contains ibogaine, which is now being investigated as a treatment for drug addiction, without withdrawal symptoms, by scientists at the University of Miami. 'We have much to learn from the animal kingdom,' says Huffman.

It is hard enough to work out what is going through the mind of a human addict, let alone that of an ape. There is some debate over whether gorillas enjoy the *T. iboga* root in the same way that African religious cults and rites use it to 'open their heads' to collapse, hallucinate and reach out to their ancestors. Pierre Henri Chanjon, a professional hunter and guide and former official guardian of the Petit Loango Reserve in southwestern Gabon, is familiar with the root-eating gorillas and believes that the apes are intelligent enough to be discriminatory in their consumption, possibly using the plant as a tonic.

Most intriguing, local people claim to have discovered the intoxicating effects of the roots by watching animals, including gorillas, go into a frenzy of fear, acting as if they were being chased by invisible objects. In Bwiti legends the pygmies are said to have found iboga, but it is possible that the pygmies themselves discovered the properties of the plant by observing wild boars digging up and eating the roots, only to go into a frenzy, jumping around and fleeing from perhaps frightening images. Similar behaviour has been reported by indigenous peoples for porcupines and gorillas who are said to be fond of the roots. A century ago, the effects of the drug were studied by French scientists. Dogs injected with ibogaine acted as if they were seeing frightening things; they would suddenly begin to bark loudly at nothing, leap backwards or desperately try to hide in a corner. Perhaps the root affects vision, making it possible to see something invisible like a Tebo, an enchanted warthog that Newt Scamander says can be found in Congo and Zaire. This sounds like something that is worth investigating by Professor Vindictus Viridian, author of that classic and influential work, *Curses and Counter-curses (Bewitch Your Friends and Befuddle Your Enemies with the Latest Revenges: Hair Loss, Jelly-Legs, Tongue-Tying and much, much more).*

Ancient potions

Within the infant rind of this weak flower
Poison hath residence and medicine power.

Shakespeare, *Romeo and Juliet*

The earliest systematic study of Snape's medicine is thought to have been conducted by the Chinese Emperor Shennung, who lived around 2700BC, and was written up in the *Pen Tsao* more than two millennia later. This herbal contains substances that have since been confirmed as effective agents, as does the 1500BC Ebers papyrus from ancient Egypt, which discusses plant extracts, minerals, animal organs and magic.

The papyrus refers to opium as an ingredient in a remedy for colic. The sticky white sap of the seed capsule of *Papaver somniferum* (*Papaver* is the Greek noun for the poppy, and its species name, *somniferum,* is from the Latin word meaning 'sleep inducing') contains 25 per cent by weight of the opium alkaloids. The poppy also features in Greek mythology, in which it was dedicated to the god of dreams, Morpheus. Representations of the twin brothers Hypnos and Thanatos (the gods of sleep and death) showed them crowned with poppies or holding the flowers, demonstrating the Greek awareness that one may never awaken from sleep induced by opium. Pliny accompanied a description of how to collect raw opium with a warning that 'taken in too large quantities is productive of sleep unto death even'.

It is, as Snape told his class, also possible to stopper death in a phial. To work out what exactly to stopper, Cleopatra used her slaves as guinea pigs. Extracts from deadly nightshade were quick but painful; the same went for henbane (its most active ingredient hyoscine, scopolamine, is now used as a sedative); strychnine was rapid but left the face contorted in death; however, asp's venom produced a speedy and tranquil demise – ideal for a queen with suicidal tendencies.

Snape also mentioned one traditional way to counter these poisons: a hard mass, such as a stone or a hairball, in the stomach of an animal such as a goat or antelope was thought to be an antidote. It was called a bezoar after the Persian for 'counter poison'. It is doubtful the bezoar could defeat toxins, though ancient people, and even monkeys and apes, did know that soils, clay and charcoal can bind toxins.

Knowledge of the properties of plants continued to accrue. In the first century, the Greek physician Pedanius Dioscorides wrote a compendium of 500 medicinal plants called *De Materia Medica* which was influential until the early Renaissance. During the Dark Ages, various potions and poultices were described in 'leechbooks', a term derived from the Anglo-Saxon *laece*, a doctor.

One well-known example, written by Bald and Cild around AD900, had some recognisable treatments with some understandable actions, at least to today's scientists. But there was probably more placebo than real pharmacology, not to mention fantastic brews. Wormwood, for example, was used with lupin and singing (nine masses!) as a 'salve against the elfin race and nocturnal goblin visitors'.

The *Liber de Proprietatibus Rerum* of Bartholomaeus Anglicus, which appeared around 1250, included information on angels and demons, winged creatures such as griffins along with descriptions of medicinal plants and the body's essences. It would take another few hundred years for the foundations of the understanding of drugs to be laid as a result of advances in chemistry.

One important contribution to the field was written in the sixteenth century by John Gerard. In the 1,392 pages of *The Herball, or generall historie of plantes* (1597), the Elizabethan herbalist and barber–surgeon assembled 2,200 woodcut images and information on a vast range of plants, some of which would be well known to Snape. Gerard mentions aconite, from *Aconitum napellus*, a poison that has been used since classical times. According to mythology, the plant sprang from the vomit of Fluffy's famous three-headed precursor, Cerberus. Its deadly properties were noted by Hecate, the goddess of witchcraft and mistress of the art of poisoning.

In his potions class, Snape reels off some of the alternative names for aconite: wolfsbane (used to kill wolves) and monkshood, a reference to how its purple sepals are fancifully shaped, one of them being in the form of a hood. Other names are less subtle: woman killer, brute killer and leopard killer. When taken by mouth, the toxic alkaloid chemicals in aconite produce a burning and tingling sensation because they interfere with the way charged sodium atoms drift across cell membranes. Eventually aconite interferes with the heartbeat and respiration. Gerard described how the tongue and lips swell up, eyes pop out, thighs stiffen and the strickens' 'wits are taken from them'.

By linking doses to the way patients responded, as a doctor would today, herbal medicine could be made more effective. This advance was reported in *An Account of the Foxglove and some of its Medical Uses*, written in 1785 by William Withering (1741–99). In his day, foxglove was used to treat dropsy, a swelling of the limbs and torso that we now know to be caused by the inadequate pumping action of the heart.

Withering's preface declared: 'The use of the Foxglove is getting abroad, and it is better the world should derive some instruction, however imperfect, from my experience, than that the lives of men should be hazarded by its unguarded exhibition, or that a medicine of so much efficacy should be condemned and rejected as dangerous and unmanageable.'

During his potions class, Snape mentions another plant that can be dangerous in large quantities. Wormwood is the bitter ingredient of the pale green liqueur absinthe, the legendary nineteenth-century alchoholic beverage enjoyed by the likes of Gauguin, van Gogh and Baudelaire. Consuming it in excess was linked with brain damage and even death. Wormwood contains thujone, which stimulates the nervous system, and has been used for thousands of years in a range of remedies, including, somewhat unsurprisingly, to treat worms. However, Snape also describes mixing wormwood with asphodel, a member of the lily family with clusters of white flowers that was associated by Greek legend with the underworld. The result, he claimed, was a sleeping potion called the Draught of Living Death.

Screaming mandrakes

. . . shrieks like mandrakes, torn out of the earth . . .

Shakespeare, *Romeo and Juliet*

Harry learned much about the strange mandrake plant during herbology classes with Professor Sprout, a little witch. Mandrakes were used, for example, to make a restorative draught to counter petrification, a fate suffered by Hermione Granger, Colin Creevey and Mrs Norris, the caretaker's cat.

Harry's copies of *Magical Draughts and Potions* and *One Thousand Magical Herbs and Fungi* probably contain copious references to mandrake's extraordinary reputation. The root of the plant looks like a small, muddy and grotesquely ugly baby, according to J.K. Rowling. When they enter adolescence, mandrakes supposedly become moody and secretive, just like their human equivalents.

In the Muggle world, the plant can be found in countries bordering the Mediterranean. It has a Y-shaped root and was indeed thought in ancient times to resemble the human form. This was significant because, according to the prevalent 'doctrine of signatures', the properties of a plant could be predicted from its appearance. The saffron plant, being yellow, was thought to treat jaundice. Cyclamen plants have leaves that resemble the human ear, and were used to treat hearing disorders. Mandrake, which has roots that resemble the shape of the body, was linked with fertility and prescribed for a wide variety of ailments.

Because mandrake was so highly prized, herbalists invented all kinds of legends to deter people from digging up the roots. In Harry Potter, we learn that the cry of the mandrake is fatal (the herbology class wore earmuffs). The plant also puts up a fierce struggle if uprooted, flailing its fists and gnashing its teeth. All this has a historical precedent. Theophrastus (372–

287BC), who described the mandrake in his *Enquiry into Plants*, recommended special precautions, which involved tracing a circle around the plant with a sword. The use of a dog to dig up the root was recommended from at least the first century AD and perhaps earlier. One reason for employing a dog for this job, apparently, was to avoid the terrible scream the plant would make when pulled from the ground. In medieval Europe, this howl was thought to be deadly.

The mandrake could indeed kill: fermented root was used by Renaissance poisoners, such as Cesare Borgia. But at lower doses it found other uses. The *Grete Herball*, published by Peter Treveris in 1526, refers to how 'if mandragora be taken out of measure by and by slepe ensueth and a great lousing of the streynghe with a forgetfulness'. In Shakespeare's *Othello*, mandragora is called a 'drowsy syrup'.

The plant, which was used as a reviving potion in the second Harry Potter book, contains powerful alkaloid toxins that produce a dry mouth, double vision, hallucinations, sleep and even coma. In earlier times, mandrake could have put people into an altered state of consciousness for surgery, to cope with childbirth, help them to sleep, or even give them a break from traumatic experiences.

The potions live on

Among the many and varied concoctions of strange, bizarre and downright disgusting ingredients used in ancient times, a great number are still in use by wizards and witches in hospitals. Today, the legacy of herbal medicine lives on in a range of highly purified, single-substance drugs. While herbalists employ entire plants, drug companies prefer to isolate the active ingredient to cut the risk of side effects and boost potency. We still use the analgesics morphine and codeine from opium, and taxol from the Pacific yew to treat cancer.

Other cancer treatments include vinblastine and vincristine from the rosy periwinkle, a pretty little flower from Madagascar. The alkaloid reserpine, from the snakeroot plant *Rauwolfia serpentina*, a shrub with pinkish-white flowers, lowers blood pressure.

The bark of the cinchona tree, used in the Inca empire to treat fever, contained bitter quinine, and was effective against the malaria parasite (until resistance set in). Scientists now know that foxgloves, such as *Digitalis purpurea* and *Digitalis lanata*, act on heart muscle to help treat heart failure. *D. lanata* contains digoxin, a cardiac glycoside, which increases the contractility of heart muscle. Even the ergot that terrorised Salem has a good side: it can quicken labour and one of its alkaloids, ergometrine, can stem bleeding after birth by contracting blood vessels.

Willow bark and queen of the meadow, *Filipendula ulmaria*, were used for a long time in folk medicine to treat pain and fevers. They are both sources of salicin, the precursor of modern aspirin. Over the past two decades this mild painkiller has been found to reduce the risk of strokes, heart attacks and bowel and colon cancer, protect nerves and inhibit a protein that helps the AIDS virus to multiply.

Like many herbal drugs, aspirin is what pharmacologists call a 'dirty' drug because it binds to many molecular targets in the body rather than just one. The best known are the so-called cyclooxygenase enzymes COX1 and COX2. At the lowest dose of one baby aspirin a day, COX1 is inhibited in blood platelets, which prevents their sticking to vessel walls – hence the drug's effectiveness in warding off heart disease. Two tablets inhibit both COX1 and COX2 in most tissues, reducing inflammation, lowering fever and reducing pain. At the high dose of 5 to 8 grams a day, used to control rheumatic fever and rheumatoid arthritis, there is enough aspirin to inhibit IKK-beta, an enzyme that stops insulin from working.

This discovery, made by Steve Shoelson of Harvard Medical School in Boston and colleagues, has suggested a new way to treat diabetes. More recently, another American group reported that aspirin can fight human cytomegalovirus, which can cause birth defects and kill people with weak immune systems. Aspirin alone still has much to teach us. Even today, Snape's potions could pave the way towards a range of new drugs.

Ethnobotany

Many traditional medicines are now being studied by scientists because there is often more than a grain of truth to what the ancients claimed. This field is called ethnobotany, a term coined by the American botanist John Harshberger in 1895 to describe studies of 'plants used by primitive and aboriginal people' and not, as some seem to think today, the hunt for drugs in natural products using high technology. There is plenty left to investigate. Since the start of modern pharmacology, only around 1 per cent of the 250,000 species of flowering plants have been exhaustively studied for their medical uses, leaving limitless pharmacological potential.

Ethnobotanists remain fascinated by the poison dart of a Shipibo hunter or the medicinal plants used by Tahitian healers. They have studied whether tea made from a lichen that Piaroa Indians call 'iguana toe' really can cure urinary tract infection or even gonorrhoea. They have asked why some Indians prefer the antibiotic properties of honey of stingless bees over honey from killer bees.

However, it has proved hard work isolating the active molecules of plants cited in old wives' tales, folk remedies and the poultices of local healers and turning this wisdom into prescription drugs. This is underlined by an effort born in the

1950s, when scientists discovered plant-derived vinblastine and vincristine as a treatment for leukaemia. The US National Cancer Institute and Department of Agriculture subsequently began screening organisms, whether plant, marine, or microbial, for useful drugs

Since 1960 tens of thousands of samples have been stored at the Natural Products Repository in Frederick, Maryland. Each year scientists test about 20,000 extracts, but only 2 per cent have shown activity against cancer or AIDS, for example. Since 1960, only seven plant-derived anticancer drugs have received Food and Drug Administration approval for commercial production. Since 1986, over 40,000 plant samples have been screened, but so far only five chemicals showing significant activity against AIDS have been isolated. At the time of writing, three were in preclinical development, isolated from an Australian shrub and trees in Sarawak, Malaysia.

Meanwhile, the technology to screen active molecules has advanced to the point where it is possible to screen many tens of thousands of samples in a day, regardless of what the local witchdoctor prefers to use. However, ethnobotanists remain adamant that their field offers a more efficient way to find drugs. Michael Balick, testing National Cancer Institute samples for potential Aids drugs, found that a handful of 'powerful plants' from a village healer in Belize gave four times as many potential candidates as a random collection.

Screening technology will never make ethnobotany redundant, added Mark Plotkin, ethnobotanist and author of *Tales of a Shaman's Apprentice*. 'The Amazon is home to 80,000 flowering plants. What are the odds of re-creating that potion using four plants – remember, it is not just which four plants, but also which part of the plant (flowers? inner bark? and so on) *plus* method of preparation *plus* dosage. It simply *cannot* be duplicated with a random screen. Not ever!'

The end of ethnobotany

Even in the wizarding world, some species are under threat: the Golden Snidget, a little bird, was driven to near extinction by its use in early Quidditch games and has become a protected species. Players now use the Golden Snitch instead, after the mechanical replacement was developed by the wizard Bowman Wright.

The Muggle world has not been so fortunate. John Riddle (no relation to Voldemort) of North Carolina State cites the fate of a form of herbal birth control that created much of the wealth of the Greek city-state of Cyrene on the coast of what is now Libya. The Cyrenians collected and exported the sap of a plant that the Greeks called silphion and the Romans silphium, which was probably a giant fennel of the genus *Ferula*.

The ancient physician Soranus refers to how 'Cyreniac juice' could be drunk to prevent conception or induce an abortion. Like other plants, it probably contained oestrogen-like compounds that alter the subtle balance of hormones needed for conception and maintenance of pregnancy. It was in great demand and the Roman naturalist Pliny the Elder mentions that silphium cost more than its weight in silver. By the fourth century AD, however, the supply had run out, apparently harvested to extinction. There were other herbs that had the same effect but, intriguingly, Riddle believes that the knowledge of how to use them was also lost: women who possessed the secrets of fertility control were burned as witches in the Middle Ages, being first in line to be blamed for sterility, miscarriages, impotence and stillbirths.

Today, forests in Africa, Indonesia and Malaysia are falling swiftly to loggers. Well over half the world's plant species are to be found in tropical rainforests which represent, presumably, half of the plants with medical potential. Day by day they are becoming harder to find. One healer in Belize used to gather medicinal plants a ten-minute walk away from his house in

1940. Almost half a century later he had to walk seventy-five minutes to find them.

Rainforests cover barely 7 per cent of the earth's surface, and they are shrinking, by some estimates, at up to 100 acres a minute. Many plants are now endangered as a result. The International Union for the Conservation of Nature has placed more than 22,000 seed plants on its Red List, suggesting that about 10 per cent of the world's flora is threatened with extinction.

To date, we have not made much use of the 250,000 species of flowering plants. Chinese traditional medicine relies on about 5,000 of them, yet the 120 or so plant-based prescription drugs sold around the world are derived from just ninety-five species, according to Norman Farnsworth of the University of Illinois, Chicago. We will probably never know more than a fraction of what lies in the great apes' and shamans' pharmacies.

As the ingredients of Snape's potions disappear, so have the recipes to use them. Traditional knowledge built up by trial and error over thousands of years is being lost as local people 'disappear into suits and ties', according to Mark Plotkin, who is also the president of the Amazon Conservation Team, Virginia. This cultural homogenisation is mirrored by the loss of linguistic diversity, with something like half of the world's 5,400 extant spoken languages threatened with extinction.

Fortunately, some remarkable hotspots of biodiversity survive, places where future discoveries will be made. One lies off the Horn of Africa, the island home of a range of plants that sound like something imagined by J.K. Rowling. The island of Socotra (Soqotra) has remained largely untouched since prehistoric times. Of its 850 plant species, around one-third are unique. Its weird vegetation makes Socotra the tenth richest island in the world in terms of endemic plant species, according to the World Conservation Monitoring Center.

Perhaps the most strikingly primitive plant is the dragon's

blood tree (*Dracaena cinnabari*), sometimes called the 'inside-out umbrella tree' because of its odd shape. Legend has it that the tree sprung up from congealed blood shed by a dragon and an elephant as they fought to the death. Cinnabar, the crimson red resin from its leaves and bark, was prized in the ancient world, where it was used as a pigment, for treating dysentery and burns, and fastening loose teeth and freshening breath.

Socotra also sports examples of gigantism, a curious phenomenon of island evolution. When the island was still part of Africa, any broad-trunked trees would likely have been destroyed by elephants, rhinoceroses and other larger herbivores. When Socotra broke away some ten million years ago, it left the herbivores behind, providing an ecological niche into which plants could grow and grow, and then grow some more. The most startling example of gigantism is the cucumber tree (*Dendrosicyos socotrana*, which grows up to 4 metres (around 13 feet) tall – a far cry from the climbing plants or shrubs that typically feature in the cucumber family. Socotra sounds like a promising, although slower, alternative to enlarging plants by genetic wizardry that I discussed earlier.

CHAPTER 12

The origins of witchcraft

'Tis now the very witching time of night,
When churchyards yawn and hell itself breathes out
Contagion to this world . . .

Shakespeare, *Hamlet*

He defeated the dark wizard Grindelwald. He was asked to be Minister of Magic. He discovered twelve uses for dragon blood. He was awarded the Order of Merlin, First Class. He is now Chief Warlock, Supreme Mugwump and a member of the International Confederation of Wizards. I am, of course, referring to Albus Dumbledore, headmaster of Hogwarts School of Witchcraft and Wizardry.

Using an authoritative source – Chocolate Frog cards – we also know that Dumbledore is now regarded as a world-class sorcerer, the greatest wizard of modern times, who once worked with the alchemist Nicolas Flamel. But should we believe everything that we read about him? Did wizards ever really wander around in cloaks and pointed hats and brandish wands? Did Dumbledore and his kind really exist? Evidence from archaeology, anthropology and psychology can provide some intriguing answers.

To complement the efforts of these various 'ologists, historians have also sought out records of wizardry and witchcraft. Indeed, in one Oxford museum it is possible to find a silvered and stoppered bottle discovered by a Muggle historian that allegedly contains a witch.

In most of this book I have loosely equated witchcraft with magic, but it turns out that the concept of witchcraft is a little more slippery than that. This is hardly surprising, given its history – which has everything to do with mistrust, ill-fortune and prejudice and little to do with black magic, sorcery and the occult.

Today, there are those who quite happily admit to being a witch, though usually one of the benign white variety. However, when fear of witches was at its peak, such an admission would have led to a gruesome death, for it implied that you had a pact with the Devil. You were a member of a conspiratorial cult that served Satan to subvert human society and God's world. You were tapping supernatural forces for wicked ends. In sixteenth- and seventeenth-century Europe it is estimated that 50,000 so-called witches were burned at the stake. Intriguingly, Muggles think they can explain the reasons for this extraordinary death toll, and it is not just a matter of religion, social forces and psychology. A change in climate also contributed.

The real Dumbledore

The wise and benevolent headmaster of Hogwarts, Albus Dumbledore, is seen by some as being reminiscent of Merlin from the King Arthur legends. But Robin Briggs, an Oxford historian who has studied witchcraft, believes that the closest parallel – perhaps even inspiration – for Dumbledore is much more recent and tangible, a historical figure. The real Dumbledore is also, in some ways, even more extraordinary.

Enter John Dee, the occultist and adviser to the queen, Elizabeth I.

The Harry Potter books reassert the 'old tradition of thinking of magic and witchcraft as all part of some secret knowledge for special people', says Briggs. In Britain, there was a tradition of elevating men who captured the public imagination into wizards. One was Roger Bacon, the monk, scholar and scientific pioneer who supposedly made a wonderful head of brass that spoke three times: 'Time is . . . time was . . . time's past.' Other wizards include Cardinal Wolsey, the statesman who dominated Henry VIII's foreign and domestic policies until he failed to get papal consent to annul the king's marriage; Owen Glendower, the Welsh chieftain who led a revolt against Henry IV; Sir Walter Raleigh, the courtier, explorer and writer who introduced tobacco and potatoes to England; and Oliver Cromwell, the victorious leader of the parliamentary army in the English Civil War.

Of all those people who claimed throughout history to have tapped into mysterious hidden knowledge, John Dee is one of the most significant. When it came to the British Empire and understanding how the navy was the 'master key' to British military strength, Dee was a visionary. He coined the word Britannia, applied geometry to navigation, and was an important figure in Tudor geography who charted the Northeast and Northwest Passages. A great astronomer and mathematician, Dee also studied alchemy, crystal-gazing and astrology. He was one of the first modern scientists and one of the last serious occultists. He loomed so large in Elizabethan England that some scholars think Dee served as model for Marlowe's Dr Faustus and Shakespeare's Prospero in *The Tempest*.

Dee was born on 13 July 1527, when the differences between astronomy, astrology, alchemy and magic were indistinct. Even the government was interested in alchemy at that time. By one 1574 account, Sir Thomas Smith, secretary of state, had a 'great Opinion of it'. This was also the period when

Protestant theologians would debate the difference between a prayer and a spell, concluding that words had no power in themselves, unless heard by the Almighty.

To Dee, magic was not about superstition but about the study of hidden forces that ruled nature, both spiritual and physical. His magical reputation began when he was nineteen. A reader in Greek at Trinity College Cambridge, Dee arranged 'special effects' for his production of Aristophanes' play *Peace* to make a giant dung beetle fly. From this time onwards, Dee was plagued by accusations of sorcery, or as he put it in his famous *Mathematicall Praeface to Euclid's Elements*, 'a companion of the Hellhounds, and a Caller, and a Conjurer of wicked and damned spirits'.

After the accession of Queen Mary I in 1553, Dee was accused of calculating, conjuring and witchcraft, charges that referred to his work in astrology and mathematics. Informers claimed he had 'endeavoured by enchantments to destroy Queen Mary'. However, after Mary died in November 1558 and was succeeded by Elizabeth, Dee's fortunes began to improve.

Lord Dudley, one of the new queen's favourites, asked him to pick a 'propitious day' for her coronation – he selected 15 January 1559. Five years later, after a continental tour, Dee presented his work *Monas Hieroglyphica* to Elizabeth who, intrigued, would end up studying it at his side, like an apprentice with Dee as her sorcerer. Just as the Minister for Magic, Cornelius Fudge, would send daily owls to Hogwarts seeking Dumbledore's opinion, so Her Majesty would seek Dee's advice on matters astrological and alchemical – for instance when a blazing star appeared in the sky, or how to deal with the discovery of an effigy of Her Majesty stuck with pig bristles. The queen was receptive to his ideas. After, all, she regarded her own position as magical in some ways: Elizabeth was an enthusiastic practitioner of the 'royal touch' to cure epilepsy and scrofula.

In 1581, Dee went the way of Professor Trelawney and began to experiment with crystal-gazing and conducting his divination using a crystal ball (shew stone) or a clear pool of water. He deluded himself into believing that he could communicate with the spirit world. Unlike Dumbledore, however, Dee sounds somewhat gullible. In March 1582 he met Edward Kelley whom he believed to be a skryer, a spirit medium. Unfortunately for Dee, Kelley was also a conman. In a peculiar twist, Dee also concluded that Kelley (who called himself Edward Talbot when they initially met) was a cosener (fraud). However, they were reconciled later that year.

This period of Dee's life would see attitudes towards astrologers shift. The conjunction of Saturn and Jupiter in 1583 was believed to mark the onset of an era of catastrophic events that would climax in the *annus horribilis* of 1588. Presumably the astrologers had got the Continental report, since the year saw the assassination of Henri de Lorraine, duc de Guise, who was notorious for a massacre of Protestants in 1572; the quelling of plots against the queen; the first English colony in the New World; and, most significant of all, the defeat of the Spanish Armada. It was one of the most glorious in English history.

Despite these emerging doubts about divination, Dee believed that Kelley was channelling messages from angels, complete with 'Enochian language' that God had supposedly taught to Adam, and a large cast of spiritual characters. They would also work together on the philosopher's stone, an effort that would bring Kelley great fame along with his knowledge of necromancy – supposedly conjuring up the dead, for instance to obtain knowledge of the future.

From available evidence, Dee did indeed look like Dumbledore. A description was provided by Elias Ashmole, who in 1672 visited Mortlake to interview Goodwife Faldo, the only person left in the village who had encountered Dee. She remembered that he was handsome, tall and slender, with a fair complexion, dressed in a gown with hanging sleeves, and had a long, pointed beard that, in old age, was white. Children were frightened by Dee, who was 'accounted a conjurer'. By comparison, Dumbledore was tall, thin and very old. His hair was silvery, and his beard long enough to tuck into his belt. He wore half-moon glasses, perched on a long crooked nose.

After a six-year peripatetic odyssey with Kelley from England to Poland to Prague and southern Bohemia, Dee returned to Mortlake, Surrey, in December 1589. His relationship with Kelley had ended two years earlier, when Kelley – who lusted after Dee's young wife and loathed his own – engineered a wife-swapping operation 'at the angels' command' that left Dee's wife pregnant.

Dee eventually died in penury under further suspicion of witchcraft. As if to underline how low his star had fallen, we do not know the exact date of his death, except that it took place between December 1608 and March 1609, just before a telescope was turned to the sky by one of the founders of modern science, Galileo.

Dee had buried his books of mysteries in the fields around Mortlake before he died. His papers were excavated a few years later by the antiquarian Sir Robert Cotton. Transcripts of the decaying documents made by the scholar Meric Casaubon were published in 1659 as *True and Faithful Relation of What Passed for Many Years between Dr John Dee and Some Spirits*. The accompanying image made Dee, by one account, look like the Merlin of Mortlake.

Casaubon himself thought Dee was deluded. In the nineteenth century Dee was dismissed as a charlatan and his name lived on only in the murky myths of occultists: he has been credited with being the founder of the Rosicrucian movement, a secret brotherhood supposedly handing down esoteric wisdom. In fact, Dee realised only too well that the knowledge of his day was unequal to the task of revealing the secrets of the cosmos. He once confessed to Rudolf II, the Holy Roman

Emperor and a great supporter of alchemy: 'I found [at length] that neither any man living, nor any book I could yet meet withal, was able to teach me those truths I desired and longed for.' Rudolf II probably realised it too. He imprisoned Kelley because his transmutations were, of course, fraudulent.

Wizard gear

Lift up the lid of Harry Potter's large wooden trunk and you will find the standard paraphernalia of witchcraft: pointed hat, black robes, a wand and assorted spellbooks. These items are not unique to Hogwarts School of Witchcraft and Wizardry. Many similar objects have been made by Muggles over a period of thousands of years which were undoubtedly connected with magic or superior knowledge.

When Harry was invited to Hogwarts, he was also asked to bring a pewter cauldron. This significant piece of hardware has been in action for a long time. In Greek mythology, the sorceress Medea had one to restore lost youth (she cut an old ram into pieces and tossed them into the cauldron, and a young lamb stepped out). A similar cauldron appears in the Celtic (Welsh) stories of the Mabinogion, which feature the giant, Bran. The cauldron was the symbol of plenty associated with the Celtic (Irish) deity Daghda.

In Britain, bronze cauldrons date back to around 1300BC. Then came iron cauldrons. By the Middle Ages, pewter appeared – although pewter cauldrons of the kind sought by Harry are not known to Muggle experts, perhaps because their high lead content would have meant that, if heated over a fire, they would disintegrate with toxic effects.

Cauldrons were very much a feature of real-world rituals, not just as accessories to feasts but as offerings to gods, when precious objects were dropped into watery places from which they could not be retrieved. Oxford's Ashmolean Museum has one example, deposited in Shipton-on-Cherwell, Oxfordshire, about 1300BC. Cauldrons dating back to around 900BC were found in a Bronze Age deposit in Dowris, Ireland. And some time between 800 and 600BC, cauldrons were tossed into a sacred lake, Llyn Fawr, in what is now Glamorgan.

Perhaps the most stunning example of a cauldron with magical associations was discovered by peat cutters in the Raevemose bog on 28 May 1891 near the hamlet of Gundestrup, Denmark. The find consisted of a silver bowl-shaped cauldron base containing eight curved side plates, decorated with fantastic scenes of people and mythical beasts, which had once been soldered together to form the sides of a cauldron. Measuring around 27 inches across, the 'collapsible' cauldron was made in southeastern Europe, around the Lower Danube in what was then Thracia. Dismantled, it had been dumped on dry ground more than two millennia ago and left untouched so that peat grew over it. One academic suggested that the cauldron was unmolested because it was a cult object and protected by a ritual barrier that was, as he put it, as effective as a high-voltage fence.

The Celtic- and Thracian-inspired decorations on this enigmatic bowl certainly suggest it had great significance and was used for ritual purposes: there were 'scenes of war and sacrifice; bearded deities wrestled ferocious beasts, a bare-breasted goddess stood flanked by elephants, and a commanding figure in stag antlers brandishing a ram-headed snake in one hand and a twisted neck collar in the other', says Timothy Taylor of the University of Bradford, who has studied the cauldron.

The elephants and the human figures of mixed or ambiguous gender, some blending with animals, were similar to those on artefacts elsewhere in Europe and even Asia, revealing how silversmiths had contacts stretching for thousands of miles. Perhaps the most striking image to be found on the cauldron is a figure with antlers, squatting in a posture resembling that of a low-caste sorcerer (adopted even today by yogis in India), who also appears on a seal stone found in the Indus valley of South Asia, and who bears some resemblance to the Celtic horned god of the underworld, Cernunnos. This is a remarkable testament to how similar magical traditions existed across Eurasia at that time, says Taylor. Indeed, there is even a connection with Harry Potter: his father was an Animagus who could turn into a stag.

The stag antlers that adorn the figure on the cauldron lead us to another familiar feature of magicians, wizards and wise men: their fondness for impressive hats, whether a feathered headdress, the wooden head gear of Alaskan sea hunters, the bird headdress of a Mayan ballplayer, the mitre of a bishop or the mortar board of an Oxford don. Impressive headgear is worn for good reason. Ever since our ancestors bashed one another other over the head with wooden clubs, size has been an important element in the battle for power and, in particular, the struggle for food, shelter and mates.

Dominant animals, the so-called alpha males, were usually bigger than their counterparts and, as a result, more successful at passing their genes on to the next generation. Physical size has, therefore, become associated with evolutionary success. The same holds true for humans. In the 1960s and 1970s, Thomas Gregor, an anthropologist at Vanderbilt University, lived among the Mehinaku, a tropical forest people of central Brazil. Unfortunately for shorter people, called the 'peritsi', Gregor found that the taller the man, the more girlfriends he had.

A wizard or wise man would put on a tall hat and a robe over his shoulders to boost his size and thus elevate his status. You don't have to be a psychologist to figure this out but, as ever, that has not stopped them from taking a closer look. One oft-quoted experiment was conducted in 1968 by psychologist Paul Wilson. He introduced the same unfamiliar man to five groups of students – either as a student, lecturer or professor – and found that the stranger's perceived height increased in proportion to his perceived status. American psychologists Leslie Martel and Henry Biller reported in *Stature and Stigma* (1987) that short men were thought to be less mature, less positive, less secure, less masculine, less successful, less capable, less confident, less outgoing, more inhibited, more timid, more passive – and so on.

These beliefs may become self-fulfilling prophecies, with taller people being more successful and earning more. The perceived link between height and success is so strong that people often overestimate the height of high-status individuals, such as media celebrities (Dustin Hoffman, for example, is five feet five inches tall and Madonna is just five feet four). It is enough to make one dash out to buy a pointed hat – or, if you were in charge of an ancient society, have one made with the sole objective of dazzling your peers.

Archaeologists in Germany now believe they know where the distinctive shape of wizard headgear originated, though the real thing was not made of crumpled cloth but of gold sheet that was beaten into a cone and intricately embellished with cosmic symbols. Their conclusion came from a study of odd but striking cone-shaped gold objects that date back to the Bronze Age and were discovered over many years at Schifferstadt near Speyer and Ezelsdorf near Nuremberg in Germany and at Avanton near Poitiers in France. Bowl-shaped gold skull caps were also found in Ireland and Spain.

Many theories were inspired by these finds, which date back to between 1400 and 900BC. Rather than components of suits of armour, ceremonial vases, bowls, or the top of wooden stakes that surrounded Bronze Age ceremonial sites, Sabine Gerloff of Erlangen University suggested in 1993 that these gold cones and bowls were ceremonial hats – crowns – worn by Bronze Age king–priests. There may even have been a matching gold cape. The 'Gold Cape of Mold' discovered in Wales in 1833 probably formed part of a king-priest's ceremonial dress, she says.

The king–priests were believed by people at the time to have supernatural powers. This is underlined by the decoration on the hats, which echoes the embroidery on Dumbledore's green robes. Typical of markings found on gold sheet objects of the time, the hat decoration consists of bosses and concentric circles thought to correspond to the moon and the sun: 'They showed that the wearer was in contact with the gods, also symbolised by the cone pointing to heaven, and knew the secrets of celestial movements, possibly of the future,' says Gerloff.

A 3,000-year-old 30-inch-tall Bronze Age cone of beaten gold was studied at Berlin's Museum for Pre- and Early History by Wilfried Menghin and colleagues. They found that the 1,739 sun and half-moon symbols decorating it correspond approximately to a quarter of the cycle named after the Greek astronomer Meton of Athens (*c.* 440BC). Meton noticed that 235 lunar months made up almost exactly nineteen solar years. This nineteen-year lunar cycle – the Metonic cycle – was the basis for the Greek calendar until the Julian calendar was introduced in 46BC. The discovery that the cycle was decorating these Bronze Age hats suggests a sophistication at odds with the common view that locals were primitive farmers.

Another cone, discovered near the German town of Schifferstadt in 1835, is the earliest known example of these 'calendar hats' so far found, dating back to 1400BC. The cone even had a chin strap attached to it, once again suggesting it was used as headgear. Gerloff believes that the origins of the wizard's pointed hat may date back even further, from the late third millennium onwards, to the Near East where they appear on small bronze statues, on cylinder seals, carved into hard or semi-precious stones, a form of ancient printing press, and on huge Hittite stone statues made by the inhabitants of what are now Turkey and Syria between the fourteenth and thirteenth century BC. In the later second millennium, smaller bronze statuettes topped with the distinctive headgear were exported

– possibly with religious beliefs, astronomical wisdom and other knowledge – from the Near East via Cyprus and Greece to the western Mediterranean, with two even ending up in the Baltic in Sweden.

Wizards also carried wands, of course. Harry Potter uses a supple combination of holly and phoenix feather that measures around eleven inches. The remains of similar artefacts have been found by Muggle archaeologists. In Oxford the Pitt Rivers Museum contains a beautiful example, made of bone, used by a 'devil doctor' to detect those who had used witchcraft to cause illness. Apparently he would summon spirits with the help of a live mouse.

One of the best known examples of a wand was discovered in an ancient sacred site on the rugged Gower peninsula of Wales, where the Red Witch of Paviland had been laid to rest in a ceremonial burial. The body of the Red Lady, as she came to be known, came to light in 1823 when William Buckland, Professor of Geology at Oxford University and curate at Christchurch, discovered her headless remains in Goat's Hole Cave. At first he thought the bones were of a male, perhaps a customs officer murdered by smugglers. But later that year, when he published his findings, he changed his story.

The woman was brick red because she had been dusted with ochre, one of various natural earths containing iron oxide that were used as pigments in ancient times. Buckland speculated that the 'painted lady' had serviced the needs of nearby Roman soldiers. Crucially, he believed that the human remains had to be much more modern than the bones of mammoth and woolly rhino found nearby because, as Dean of Westminster, Buckland thought that such species had not made it onto Noah's Ark.

Scattered about the Red Lady's remains were stained pieces of mammoth ivory that had been fashioned into rods. As Buckland described: 'I found forty or fifty fragments of small ivory rods . . . [also] . . . some small fragments of rings made of the same ivory . . . nearly the size and shape of a small teacup handle.' There were also a tongue-shaped piece of ivory, some periwinkle shells and part of a sheep's shoulder blade. 'The Blade Bone of Mutton gives grounds for a conjecture, which favours the Theory that she was a Dealer in Witchcraft. In Meyricks History of Cardigan . . . are some curious Stories . . . of the Magical Powers of the Shoulder Blade of a Sheep.'

Buckland subsequently referred to the Red Lady of Paviland and the ivory objects and shells as adornments, or part of a game. He was later proved wildly wrong, though to be fair to Buckland, the discovery did mark the first human fossil ever to have been recovered anywhere in the world. 'The radiocarbon dating technique was not invented until the late 1940s, so Buckland could not have known the true age of the internment,' says one of those still studying the site, Stephen Aldhouse-Green of the University of Wales College, Newport.

The remains were subsequently found by radiocarbon dating to be 26,000 years old. The Red Lady also turned out to be a young man, aged between twenty-five and thirty, who weighed about eleven stone (154 pounds) and was slightly weedy. Isotope analysis of his bones showed that he ate seafood. The rods found at the ceremonial burial by Buckland may have been wands, though another possibility is that they were intended for cutting up to create blanks for ivory beads, says Aldhouse-Green.

A wand was a symbol of power, thought to enable a wise man's will to be concentrated into an object that could be pointed so that power could be projected in a particular direction. In societies where metal-making was all-important and writing was rare or non-existent, one can imagine how metallurgical recipes were committed to memory as a 'spell' that could be repeated while waving a wand.

Having said that, it should be added that there is hardly an ancient society that did not possess objects that were longer than they were wide. These batons, staffs, sticks or rods were often wielded by men of a certain age to assert their authority for reasons too obvious to explain: in mythology, there was the staff wielded by the messenger of the gods, Hermes, which was entwined with serpents and topped by wings, called a caduceus. From Scythia, now the Ukraine, there were poles adorned with bronze rattles. At Sutton Hoo, the seventh-century site in Suffolk where a Saxon boat was found containing rich grave goods, there is a long rod with a finial (ornament) on it. The Calusari or 'horse men' of Transylvania had healing sticks.

But why don't wands, cauldrons and other artefacts from the Muggle world ever seem to be able to do anything magical? The answer is explained, of course, in the Harry Potter books. They have been neutralised by the most boring department in the Ministry of Magic, the Misuse of Muggle Artefacts Office. The civil servants who toil there have a simple mission: to prevent bewitched Muggle-made things from doing harm.

You can well imagine the kind of enchanted object that upsets the men from the ministry: the possessed teapot that starts to spray hot tea around; the pair of sugar tongs that ends up clamped on someone's nose. The Ministry of Magic has done a good job. I have never come across a malevolent teapot for example, or a flying carpet, one of the many items on the Registry of Proscribed Charmable Objects. However, I suspect that many of us have encountered the enchanted door keys that shrink to the point that they mysteriously vanish.

Real witchcraft

The word 'witch' is derived from *wicca*, a term that dates back to Anglo-Saxon times, when west German tribes settled in England from around the fifth century AD. Wicca was in turn related to *wicken*, Middle High German for 'to conjure'. Almost a millennium after the tribes arrived from Germany, the incidence of witchcraft peaked as the result of a range of factors. One was the increase in population, which made people both more vulnerable to natural disaster and more in need of scapegoats.

Another, the most significant, was the rise of Christianity. A developing combination of doctrines and beliefs from the later Middle Ages through to the early seventeenth century led many people to give credence to the notion of a super-enchanted world. As the Church gradually succeeded in imposing a more severely Christian vision, along with promoting the image of more actively engaged Devil, so belief in witchcraft was fostered. The Church inflated the importance of the occult and broadcast its attempts to fight the 'Devil' and its disciples. While Satan played a minor role in early Christianity, he became central during the Middle Ages. The most intense period of persecution of witches came in the sixteenth and seventeenth centuries, though tales about witches persisted. One eighteenth-century account describes how they transformed into hares to 'lead the hounds and huntsmen a long and fruitless chase'. As late as 1924, there was a report of a man who was imprisoned for attacking a neighbour who he believed had 'ill wished' his pig.

J.K. Rowling has herself experienced a version of these social forces today. In the United States, parents in some states asked for her books to be withdrawn from public school libraries because they supposedly glorify the occult. Ministers preached that the books were evil because of their dark and demonic undertones. Some deluded souls have even branded Rowling a witch, a striking echo of what happened hundreds of years ago.

Then, as now, people responded to the idea of Satan's followers and their demonic powers with a mixture of fascination and terror. From 1486, they even had a handy guide to help them detect and deal with witches, referred to as *Malleus Maleficarum*, or Hammer of Witches, written by Heinrich

Kramer and James (Jakob) Sprenger, two Dominican friars.

The *Malleus*, which probably features in Bathilda Bagshot's masterwork, was an encyclopedia of demonology that contained many lurid details of prevalent folk beliefs about black magic, from roasting babies to depriving men of their vital members. The book covered many pressing issues of the day:

> Whether the Belief that there are such Beings as Witches is so Essential a Part of the Catholic Faith that Obstinacy to maintain the Opposite Opinion manifestly savours of Heresy.
>
> Whether Children can be Generated by Incubi.
>
> What is the Source of the Increase of Works of Witchcraft?
>
> How, as it were, they Deprive Man of his Virile Member.
>
> Of the Manner whereby they Change Men into the Shapes of Beasts.
>
> How Devils may enter the Human Body and the Head without doing any Hurt, when they cause such Metamorphosis by Means of Prestidigitation.
>
> How they Raise and Stir up Hailstorms and Tempests, and Cause Lightning to Blast both Men and Beasts.

The broomstick makes a fleeting appearance in its notes on witch transportation: 'They take the unguent which, as we have said, they make at the devil's instruction from the limbs of children, particularly of those whom they have killed before baptism, and anoint with it a chair or a broomstick; whereupon they are immediately carried up into the air.'

The cauldron was a handy utensil for witches who had an

unhealthy taste for human flesh: one captive witch described how:

> We set our snares chiefly for unbaptised children . . . and with our spells we kill them in their cradles or even when they are sleeping by their parents' side, in such a way that they afterwards are thought to have been overlain or to have died some other natural death.
>
> Then we secretly take them from their graves, and cook them in a cauldron, until the whole flesh comes away from the bones to make a soup which may easily be drunk. Of the more solid matter we make an unguent which is of virtue to help us in our arts and pleasures and our transportations; and with the liquid we fill a flask or skin, whoever drinks from which, with the addition of a few other ceremonies, immediately acquires much knowledge and becomes a leader in our sect.

The *Malleus* reveals how these extraordinary confessions were obtained. Suspects were tortured after they had been stripped naked: 'This stripping is lest some means of witchcraft may have been sewed into the clothing – such as often, taught by the Devil, they prepare from the bodies of unbaptised infants, [murdered] that they may forfeit salvation.' (For those concerned that such interrogation methods are a tad unfair, the *Malleus* points out that if a witch confesses under the torture, she must of course 'certify that it was not due alone to the force of the torture'.)

The book appeared in many editions and had serious credentials: the *Malleus* had been approved by the theological faculty of the University of Cologne. In 1484 the authors had persuaded Pope Innocent VIII to issue a bull, *Summis Desiderantes*, which ordered that witches be purged. He declared (this gives a vivid glimpse of the tenor of the times, though it would do no harm to skip this part):

> Many persons of both sexes, heedless of their own salvation and forsaking the catholic faith, give themselves over to devils male and female, and by their incantations, charms and conjurings, and by other abominable superstitions and sortileges, offences, crimes, and misdeeds, ruin and cause to perish the offspring of women, the foal of animals, the products of the earth, the grapes of vines, and the fruits of trees, as well as men and women, cattle and flocks and herds and animals of every kind, vineyards also and orchards, meadows, pastures, harvests, grains and other fruits of the earth; that they afflict and torture with dire pains and anguish, both internal and external, these men, women, cattle, flocks, herds, and animals, and hinder men from begetting and women from conceiving, and prevent all consummation of marriage; that, moreover, they deny with sacrilegious lips the faith they received in holy baptism; and that, at the instigation of the enemy of mankind, they do not fear to commit and perpetrate many other abominable offences and crimes, at the risk of their own souls, to the insult of the divine majesty and to the pernicious example and scandal of multitudes.

Witches, in short, were the ultimate scapegoats.

The real witches

A modern witch-hunt has been carried out from All Souls College, Oxford, by Robin Briggs. The historian has devoted years of research and effort to finding out more about real witches, mostly by a meticulous examination of trial records of witchcraft cases in France, Scandinavia, Eastern Europe, Germany and New England. He could, for example, seek evidence of the suggestion by the nineteenth-century French scholar

Jules Michelet in *La Sorcière* that the witches of Christian Europe were in fact practitioners of an older religion.

No Satanic religion or systematic belief in diabolical witchcraft has yet been uncovered by Briggs, a finding that echoes what French sociologist Émile Durkheim once declared: magic serves the individual, while religion serves the group. A religious practitioner has a congregation, while a magical practitioner has a clientele: 'There is no church of magic.'

Hundreds of years ago, the stereotypical 'witch' was an amalgamation of various ancient and persistent folk beliefs: nightflight, spells, charms, weather magic, hidden powers and lines of force. There were some men and women called cunning folk (from the Old Norse *cunna*, 'to know'), who believed and practised all the above (at great risk) and were in a sense the same as witches and wizards. They had an ambiguous reputation among the hoi polloi and even helped to persecute the witches, arguing that they fundamentally differed from them. In modern parlance, they were white witches, while true witchcraft was black magic performed by those who had made a pact with the devil.

The details differed so much in the witch trials studied by Briggs that they seem to reflect less a systematic dark religion than a widespread fear of occult powers and an obsession with the secret enemies who wielded them. According to the society that created and hunted them hundreds of years ago, witches supposedly adored Satan. They were assumed, like heretics before them, to meet in secret at night, to carry out blasphemous parodies of Christian worship, and to indulge in orgies. The rare witch-hunts of the later seventeenth century saw the use of torture to extract confessions to crimes along the lines of these themes over and over again, creating the illusion of a secret society that worshipped Satan.

Some modern scholars have argued that witch persecution had very much to do with mass hysteria, or misogyny, primitive sexist attitudes and how women and midwives bore the brunt of blame for the high infant death rate in those days. Others maintain that downtrodden women resorted to magic to empower themselves in a male-dominated world. But the modern idea of a 'typical witch', a wizened woman with steaming cauldron, broomstick and warty nose, does not correspond to the majority of cases studied by Briggs. He found that witches were not universally, nor even overwhelmingly, female. Across Europe, one quarter of those convicted of witchcraft between the years of 1400 and 1700 were men, but this ratio varied from place to place: in England, there were unusually few men prosecuted under the witchcraft act, so few that they were sometimes called witches; the ratio was almost 50/50 within the jurisdiction of the Parlement of Paris, while in Iceland or Estonia men formed a majority of the accused. Witches tended to be middle-aged, rather than wizened hags. Most witches were taken to court by their neighbours rather than tracked down by fanatical 'witch-hunters'.

Once again, if witchcraft was the religious inverse of Christianity, with its own equal but opposite culture of rituals, it is a puzzle why there was so much local variation in the ratio of wizards and witches and why there were only 50,000 or so victims of continental witch-hunts over the course of more than a century, far fewer than the toll from religious strife or ordinary warfare during the same period.

Bad neighbours make good witches

People in pre-industrial society had a biologically conditioned fear of malevolent magic. In his book *Witches and Neighbours*, Robin Briggs explains how, when faced with incomprehensible and uncontrollable misfortune, whether disease, stillbirth or crop failure caused by climate change, people 'projected' the evil in their community on to an individual who was disliked for some pre-existing reason. Witch candidates were those who were envied, or unsociable, or simply had a face that did not fit. Who were these people? Annoying neighbours, of course. Many such people were accused of witchcraft, and a few probably believed it themselves, according to Briggs.

This fear of 'neighbour-witches' was probably ancient and rested on the most primitive of spatial and temporal coincidences: when you suffer misfortune, blame those who are nearby. Briggs has studied the suspicions, threats and sorcery at work around four centuries ago in villages and hamlets across Lorraine, France, to re-create the mental universe prevalent at that time. His enormous collection of surviving court records from around 380 witchcraft trials, nearly all of them in archives in Nancy, gives a vivid feel for the enchanted world of the period. People of the time did believe that sorcery had its good side. To counter magic and ward off witches, they roasted young dogs with nine kinds of herbs, buried bottles with specific ingredients under thresholds or in fields, and so on. There were other beneficial uses of sorcery: curing illnesses and identifying witches top the list, followed by a hodgepodge of practices to defend animals against wolves and other dangers, protect crops and houses, secure love, identify future partners, find buried treasure or stolen goods, protect oneself against swords and bullets, summon luck when gambling or hunting, obtain favour from the powerful and guarantee contraception.

Witches were believed to put the selfsame magic to harmful use and one consistent theme that often emerged at witch trials was how they caused sickness. In his sample of about 380 accused witches Briggs found six individuals who might be classified as witch-doctors, and another twenty whose status seems that of semi-professional healers.

Doctors treating Jennon Barthelemin around 1615 diagnosed that she was a victim of witchcraft because she had a lump on her thigh with marks 'like five claws'. A neighbour, Babelon Voirin, was suspected of summoning the lump with sorcery. Voirin knew full well that she was likely to be blamed. When Barthelemin had sent a servant to ask her for milk, the request was refused because 'if she gave it, people would say she had given the sickness'. After another convicted witch claimed Babelon was an accomplice, Babelon was tried. Another witness in this trial, Nicolas Simeon, said his wife had died also believing herself bewitched by Babelon; her sickness was unnatural – the product of witchcraft – because she was emaciated and '*frénétique*'.

The relationship between witchcraft and healing was emphasised in many confessions when the witches said the devil had given them powers to heal as well as to harm. 'This, on the face of it, seems an absurd and illogical thing for him to do,' comments Briggs. The ambivalence about this power is highlighted by the case of Nicole le Mercier, who was thought to have bewitched Vaultrin Jeandel's horse. Rather than fear the power of le Mercier the witch, he told his wife he would kill le Mercier unless she healed his nag.

The horse was given a spiced drink consisting of a pinch of pepper, five roots of ginger and wine. Le Mercier then thrust her arm down the horse's throat several times; when Jeandel's wife became worried, she replied 'that she had not walked three hundred leagues on land without learning a few things'. At her trial it emerged that she knew how to use onion seed to make a man languish for three years. Other witnesses alleged that Le Mercier had boasted that she could poison a wound herself so that the victim could be cured only by her hand. On another occasion she said 'that if she saw a man urinate, she could dry him up and make him emaciated, by taking up the earth on which the urine had fallen'. This is an example of what anthropologists call contagious magic.

The evidence of another accused witch, Jennon Villemin, highlights the then popular notion that a witch had to put something of herself into the spell and, as a result, became vulnerable to retaliation. Villemin told one of her patients that whoever had caused the sickness was in the cauldron. She boiled a mixture of herbs and wine until alight, then cried out 'Witch, witch, you are going to be very ill.' The ritual supposedly made the witch responsible suffer terrible agonies. At her trial Villemin was sentenced to whipping and banishment.

Books of spells and occult knowledge were as common as at Hogwarts. On trial in 1593 was Nicolas Noel le Bragard from Nancy, a wizard who healed with ointments, fumigations, herbal baths and soups, and was a 'stroker' (one who believed he could draw out the evil from within the body with his hands). When questioned about magical texts found in his possession, Bragard referred to 'a book containing various recipes, such as to find lost property, to have oneself loved by and to enjoy women and others he did not well remember, which made him envious to acquire that science. To that effect he tore nine or ten pages out of the book, and after this had given him the means and science to do these things, the desire grew on him to know more.'

Bragard provides another glimpse of witchcraft. As one would expect, it did not work. To bewitch and 'enjoy' a woman he fancied, named Claudine des Prunes, Bragard muttered a spell that began, '*Vous incognuz, je conjure et confirme sur vous, o vous tous grands princes d'enfer, Astarothe*.' It failed to do anything. His next spell was to obtain the favour and friendship of a seigneur, while another used nails to cure toothache.

Despite many disappointments, Bragard still believed that he needed what Briggs called a 'firewall' to protect himself from the less-than-impressive results of his magic. It worked something like Harry Potter's Shield Charm, which casts a temporary wall to deflect curses. Bragard described how 'he had drawn a circle with chalk by whose means, and with the words in a spell in a little book he showed us, he intended to make the Sibyl appear so that he could hear from her about the matters he wanted to know . . .' Unfortunately, he needed a young child to make this spell work and he was not able to obtain one. Again, his magic did not deliver. And although one spell to find some stolen money did supposedly work, his other attempts at treasure hunting failed.

Before he was tortured, Bragard was asked whether he had harmed people he had tried to treat. He responded 'that he did not know if he had made them ill, that he had not given them any powder or poison, but he feared that the imprecations he had uttered, accompanied by certain invocations of evil spirits following the spells and lessons in his books, by which he prayed that misfortune might befall all his enemies, might have produced such effects'. Ultimately he confessed to using spells to make his clients ill, so that they would consult him and buy drink for him. This was sufficient to send Bragard to be burned at the stake, along with his books and other magical materials.

There are numerous references to both flying and indeed the use of broomsticks for the purpose, in the confessions made by the accused and contemporary paintings show witches on broomsticks. As referred to in the *Malleus*, witches quite often said they had used grease of some kind to change shape or to fly, which fits in with what we know of hallucinogens.

The broomstick appears in the evidence of Appoline Belz, the forty-year-old wife of a miner tried in 1580. The wife of Pierre Quenault, supervisor in the mines, thought that Belz had made her ill. As a sign of the strange ambivalence towards practitioners of magic, Belz was then summoned to heal her, making her a poultice of bread, honey and other ingredients for her stomach and a herbal drink. The chambermaid put a broom over the door, and when Appoline saw this she cried out in anger that it 'was to test if I was a witch'. The point of this exercise was that a broom placed in this way was supposed to identify a witch by preventing her from leaving the room, explains Briggs. All this, of course, will already be known to anyone who has visited *The Salem Witches' Institute*.

Most of the time, says Briggs, witchcraft worked as a kind of therapy: the cunning folk claimed to be able to be able to cure you in one of two ways. The most important was to identify the witch, usually by encouraging the client to produce a name, so that he could be called in to remove the spell. There was a widespread belief that only the witch or the devil could bring relief where the malady was their doing. The second technique was counter-magic, which might send the spell back to the giver, or transfer it to another human or animal. 'The volume of this informal witchcraft activity far exceeded that of legal persecution, and had a far longer history.'

Weather and witchcraft

There is another reason for the orgy of witch hangings and burnings in the sixteenth and seventeenth centuries: climate change, according to Wolfgang Behringer, a history professor at the University of York. The surge in the execution of witches marks an attempt to rationalise a drastic decline in living standards, and to do something about it to remove uncertainty, and thus anxiety.

Behringer points out that the period 1450 to 1850 saw the Little Ice Age, when glaciers raced down alpine valleys, large lakes and rivers froze over, and crops across Europe failed. Heavy run-off and rains also damaged crops, and both cattle and human diseases spread in the damp. On 3 August 1562, in central Europe, the sky darkened at midday, and a heavy thunderstorm struck, destroying roofs and windows. The rainstorm turned into a hailstorm that flattened crops and vines and killed livestock. A year later, sixty-three women in the small, storm-ravaged German territory of Wiesensteig were burned. The thunderstorm was typical of an unusual cooling that hampered harvests, says Behringer. 'England may have been less affected, due to its maritime climate, with more moderate changes in temperature, but still we find frost fairs on the Thames, and witchcraft legislation in England and

Scotland was introduced at the same time as on the continent, around 1563.'

Perhaps Church officials, who had previously dismissed witchcraft as mere superstition, decided it was real and sufficiently widespread to use to bolster wavering spiritual doubts against a background of social, religious and political upheaval. At the same time, the poor, the starving and others looking for scapegoats found it convenient to accuse their neighbours. Farmworkers whose livelihoods were particularly vulnerable to the vagaries of climate clamoured for witch burning. Witches, even those with no supernatural powers, had an undeniable ability to bring out the worst in their fellow human beings.

The most vicious outbreaks of witch persecution coincided with the worst weather, according to Behringer: frigid winters and damp, chilly summers that truncated growing seasons, a delayed spring and an accelerated autumn. 'Witchcraft is the unique crime of the Little Ice Age,' he wrote in the journal *Climatic Change*.

J.K. Rowling seems unconvinced by this supposed witch persecution. Harry was, after all, asked to write an essay on how witch burning in the fourteenth century was pointless. Another reason that the persecution was not worthwhile, as revealed in the *Prisoner of Azkaban*, was that witches such as Wendelin the Weird carried a Flame-Freezing Charm and then at the stake pretended to cry out with pain while in fact enjoying a tickling sensation.

Witches today

Even today, some believe that Wicca is a real religion, an idea popularised by Margaret Murray in her 1921 book *The Witch Cult in Western Europe*. While Ronald Hutton of Bristol University does not set out to mock or condemn modern witchcraft, he points out that, although Murray's work marked the first attempt to study the great witch-hunt, her scholarship was flawed. Both her sources and her treatment of them were seriously defective.

Murray assumed that the trials were conducted against believers in a genuine religion then present in Europe, rather than practitioners of a black parody of Christianity, and attempted to reconstruct that religion from the testimony. 'When her material specified that witches indulged in orgiastic sexual behaviour, human sacrifice and cannibalism, she set this down as the truth . . . she did suggest that in its joyous nature, the witch religion had some superiority to Christianity,' says Hutton. 'She mangled data continually to fit her assertion that all witches operated in covens of thirteen, though it is obvious even from the limited data which she scanned that most of the accused were solitary individuals.'

Three decades later, Gerald Gardner, a retired colonial civil servant, went one step further in his book *Witchcraft Today*, and argued that the pagan religion had not been wiped out but had in fact gone into hiding and continued to exist. Inspired by Murray's imagination and dubious scholarship, Gardner embellished his own ideas with references to the Old Religion, the Horned God, fertility rites and covens. As a result of the popularity of the book and its successor, *The Meaning of Witchcraft*, many covens sprang up during the 1950s and 1960s. However, as Hutton points out, 'No known cult in the ancient world was carried on by devotees who all worshipped regularly in the nude like the witches . . . inspired by Gardner.'

Even today, witchcraft writers like to think that Wicca dates back to the days of the Druids, the order of priestly officials in pre-Roman Britain who revered the oak and mistletoe, practised human sacrifice, divination and astrology, and taught that the soul at death was transferred to another's body. However, the Romans thoughtfully obliterated most records of what the Druids actually did, aiding modern attempts to reinvent the Druids as proto-wizards.

Murray's idea of a witch religion was eventually demolished by scholars in the 1970s, according to Hutton. Sir Keith Thomas in 1971 and Norman Cohn in 1975 exposed the way she misused the evidence, for example. Witches were not practitioners of an old religion but, as Briggs has shown, neighbours who had made serious enemies. 'The only solid evidence for the origins of Wicca consists of the works of Gerald Gardner himself,' says Hutton. 'The official conversion of the British Isles to Christianity left no surviving pre-Christian religions, either in remote areas or as "underground" movements.'

However, after the death of Gardner in 1964, his mission was continued by others, some of whom claimed to have belonged to covens that were now emerging from the shadows. For readers of Harry Potter, this fiction sounds very familiar. Witches and wizards are alive and well, living in hiding. They are out there, leading quiet lives in leafy suburbs where they hone their spells and charms, well away from prying Muggle eyes.

The reason we are unfamiliar with the wizarding world is, of course, because of the strenuous efforts of the Ministry of Magic, which is charged with the job of keeping from the Muggles the fact that witches and wizards are crawling all over the country and zooming overhead on broomsticks and in enchanted cars. From my own experience, the ministry has done an excellent job.

CHAPTER 13

The philosopher's stone

Restore his years, renew him, like an eagle,
To the fifth age; make him get sons and daughters,
Young giants; as our philosophers have done . . .

Ben Jonson, *The Alchemist*

The philosopher's stone is the true scientific star of Harry Potter's first wizard adventure. Voldemort desperately seeks it so he can harness its extraordinary powers to regain his strength and rise up once again to spread dark magic around the world. Thanks to the intervention of Harry and his school chums, however, the Dark Lord does not succeed, joining a long line of individuals, from Chinese rulers to Holy Roman Emperors and many more besides, who have tried, and failed, in the age-old quest to find the elusive stone.

The 'philosopher' in question was a natural philosopher, the designation used for a scientist before the latter term was coined in 1840 by William Whewell, a polymath who was a fellow of the Royal Society in London. Because the term 'philosopher' sounds somewhat dull and worthy, the publisher of the American edition of the first Harry Potter book changed its title to *Harry Potter and the Sorcerer's Stone* when, in fact, it should have been *Harry Potter and the Scientist's Stone*.

Rather than a stone, some ancients referred to a 'tincture' that could bring about a transformation of a base metal into gold instantly, rather in the manner of what modern chemists would call a catalyst. Others talked about 'the powder'. Eventually, the object of the quest came to be known as the philosopher's stone, referring to how it was thought to be an inorganic substance, such as a salt or a mineral. However, this was no ordinary stone, but something very special, a 'stone which is not a stone'.

Voldemort's hunger for the stone is easy enough to understand, for, in addition to limitless wealth, it offers the prospect of eternal life, the hunt for which probably dates back to the onset of the human awareness of death. No wonder that the philosopher's stone was sought by the earliest civilisations.

Today, the search for the stone is undertaken by Muggles working in fields such as astrophysics and nuclear physics. They have revealed that the art of turning one chemical element into another is as old as the stars and started in earnest about a billion years after the Big Bang of creation, when clusters of matter formed embryonic galaxies. While they probe the secrets of this cosmic catalyst in the early universe, various types of biologist are exploring the microscopic worlds of the living cell to uncover hints of how to arrest the wear, tear and decay of everyday existence to prolong life.

Muggle bureaucrats, politicians and bean counters continue to put pressure on scientists to change the emphasis of their mission statement to focus less on the joys of fundamental discovery and more on the profit that can be made from technology and medicine to prolong life. Glossy brochures from pharmaceutical companies describe a vast range of anti-ageing research to realise Voldemort's dream of an elixir of life or a Methuselah pill, while a vast amount of other research attempts to turn this knowledge into marketing gold. The quest for the stone is as vital as ever.

Alchemy, the philosopher's stone and the elixir of life

That Alchemy is a pretty kind of game,
Somewhat like tricks o' the cards, to cheat a man
With charming.

Ben Jonson, *The Alchemist*

The search for the stone is associated with the obscure art of alchemy. Many today think of alchemy as a primitive form of chemistry, but there was much more to it than that: it extended far beyond proto-chemistry to the realm of the mystical and occult. Just as astrology focused on man's relationship with the stars, alchemy focused on his relationship with terrestrial nature, blending chemistry with magic. And the alchemists used chemistry as a metaphor for human relations, just as the astrologers used stars for the same purpose.

The ancient Greeks, Chinese and Indians usually referred to alchemy as 'The Art', or in terms referring to one of its key goals – change or transmutation in its broadest sense: chemical changes that could turn baser metals into gold, highly prized as a colourful and precious metal, one resistant to corrosion even after being buried for hundreds of years. Transmutation also implied physiological changes, from sickness to health: alchemists believed that they could use the stone to make an elixir that was able to turn dead tissue into living flesh. For the Chinese and Indians, transmutation could also mean passing from an earthly to a supernatural existence.

The idea of an elixir dates back to the Taoists in China, who were engaged in a quest to achieve immortality. Founded by the sixth century BC sage Lao-tzu, the blend of religion, philosophy, magic and primitive science led to all kinds of practical chemistry: clever methods of preserving the dead, as shown by the Tomb of Lady Tai, sealed in an airtight chamber with kaolin clay; an emphasis on precision and measuring;

apparatus, such as stoves, furnaces, reaction vessels and the still; and of course the belief that an elixir could in some way halt ageing, the search for which began some time around the fourth century BC. The most potent form of such a substance was thought to be a solution of corrosion-resistant metal, 'drinkable gold', presumably since the permanence of this noble metal would pass to whoever drank it.

One scholar counted more than 1,000 names for the elixir, of which gold was far from being the only ingredient. The book *Tan Chin Yao Chüeh* (Great Secrets of Alchemy) by Sun Ssu-miao (AD581 to some time after 673) describes formulas based on mercury, sulphur and arsenic, for example. Several Chinese emperors probably died as a result of being poisoned by such 'elixirs of life', according to the British historian of science, Joseph Needham. Over hundreds of years the failure of Chinese alchemists became obvious because they kept their goal consistent – the quest for the elixir – unlike their Western counterparts, who were also seeking gold. This is thought to be one reason Chinese alchemy faded away before its European equivalent. Another is that the Chinese adopted Buddhism, which offered a safer route to immortality.

Western alchemy dates back to the beginnings of the Hellenistic period, the period of classical Greek civilisation from around the death of Alexander the Great (323BC) to the defeat of Antony and Cleopatra (30BC). Bolos of Mendes, a Hellenised Egyptian who lived in the Nile Delta after 200BC wrote *Physica et Mystica* (Natural and Mystical Things), which contains obscure recipes to make gold and silver. Most end with a terse tribute to transmutation: 'One nature rejoices in another nature; one nature triumphs over another nature; one nature masters another nature.'

In Alexandria, Egypt, early alchemy thrived on the centuries of experience of craftsmen in using gold, manipulating it and imitating the metal, as most vividly illustrated by the astonishing golden artefacts stacked in the tombs of the pharaohs. It was their efforts that prompted discussion among the philosophers of the day of how base metals could be turned into gold. The Stockholm and Leyden papyri, which date to the third century AD, explain how to take an article of debased gold and then, with a mixture of iron sulphate, salt and alum, give it a surface of pure gold. The works of Zosimos of Panopolis, Egypt, who lived around AD300, suggest that alchemical theory came to focus on a 'tincture' that could bring about transformations instantly, and this came to be known as the philosopher's stone.

A semi-religious and quasi-magical set of ideas – from astrology and alchemy to numerology and other occult sciences – then took hold, which supposedly originated in Egypt at the time of Moses, inspired by the Egyptian deity Thoth. They are referred to as Hermetic writings, after Thoth's Greek equivalent, Hermes Trismegistus (the thrice great). Another ingredient of what is called Hermetism was supplied by the Kabbala, the doctrine of a secret, mystical interpretation of the Old Testament.

The magical world described in these writings was glimpsed only by the chosen few. Nicolas Flamel, one of those who supposedly gained access to the Hermetic art, appears in the first Harry Potter book. Flamel did indeed live in the fourteenth century and reportedly prepared the philosopher's stone. 'Among the many effects of the Harry Potter phenomenon has been the introduction of a fabled character from the history of chemistry to millions of readers who otherwise would probably have never heard of him,' comments Lawrence Principe of Johns Hopkins University, an expert on alchemy.

The classic tale is one of the great inspirational myths of alchemy. Flamel was born about 1330, probably in Paris, of humble origins and became a clerk and bookseller. The story goes that in a vivid dream, an angel presented a book on the Hermetic art to Flamel, and said: 'Look well at this book,

Nicolas. At first you will understand nothing in it – neither you nor any other man. But one day you will see in it that which no other man will be able to see.'

Later, a stranger entered his shop who was desperate to sell an old book for some much-needed money. Flamel immediately recognised the copper-bound tome engraved with strange figures and letters of an ancient language as the one shown to him by the angel. He was able to make out that it was written by Abraham the Jew. Flamel was familiar with the writings of the alchemists of his day and knew something of transmutation, but it still took him twenty-one years to solve the book's Hermetic mysteries.

As parts of the text were written in ancient Hebrew, Flamel's wife, Pernelle, suggested he consult a Jewish rabbi who was learned in the mystic Hebrew writings of the Kabbala. Knowing that many of the Jews forced out of France had gone to Spain, Flamel travelled there, to Santiago de Compostela, on a pilgrimage to the shrine of St James, trusting that he would meet some Jews along the way. During his return trip he met a Hebrew sage – Master Canches – who was able to shed light on the mysterious manuscript, giving him the keys with which he would eventually decipher the entire contents of the book.

Flamel returned home to his wife and, after three years of work, they succeeded. About noon on Monday, 17 January 1382, they turned half a pound of mercury into silver using a white type of philosopher's stone. Then, at five in the afternoon of 25 April 1382, they used a red variety to turn mercury into gold. Flamel and Pernelle went on to make the stone on three occasions.

Eventually it was claimed that Flamel also discovered the elusive elixir of life. The elixir did not seem to do him much good, however, for he died in 1417 or on 22 March 1418, depending on the source, having lived until the age of eighty-seven or eighty-eight. Today his gravestone can be found inside the Musée de Cluny, where it was moved after being used as a cutting board in a Parisian grocery.

However, one account claims that Flamel faked his own funeral, a contention backed by the first Harry Potter book, where Flamel and his wife did significantly better, surviving to the ripe old ages of 665 and 658 or thereabouts, after a quiet life pottering about Devon. How did they do it? Perhaps the answer lies in Flamel's best known work, *The Explication of the Hieroglyphical Figures – His Secret Booke of the Blessed Stone called the Philosopher's*, in which he uses religious imagery (supposedly carved on the charnel house that Flamel paid for at his local parish) to hint at how to make the stone. Like other alchemists, however, he kept the nature of the stone secret, and spoke of his work only in the most obscure and enigmatic language, providing little clue to what he was doing.

One theory suggests that Flamel claimed to have found the stone to cover up the source of his vast wealth through questionable business activities. Some sources indicate that Flamel had indeed become rich enough to found and endow fourteen hospitals, seven churches and three chapels in Paris alone, with more in Boulogne.

However, on closer examination, Principe found that the Flamel story does not hold up. 'In the world of alchemy, like in the world of wizardry, things are often not what they seem.' The Flamels were real enough but modern historians could find no evidence that they had ever studied alchemy; the earliest reference linking them with the philosopher's stone dated to around 1500, long after they had died. Flamel's most famous work, *The Explication of the Hieroglyphical Figures*, was published in 1612, and has been shown to have been written in the late 1500s. All the other alchemical texts attributed to Flamel were written after his death.

'Archival documents show that Flamel's wealth was not as

extravagant as the stories would have us believe, and that it was amassed not by transmutation but by prudent speculation in Parisian real estate and was augmented by the wealth Pernelle brought from previous marriages,' says Principe. Nonetheless, Flamel's story continued to be elaborated and embellished. Earlier accounts refer to his great wealth, but by the eighteenth century his life span had been boosted, too, no doubt with the help of the elusive stone.

In 1712, a traveller encountered a 'highly learned dervish in Asia Minor' who had just met the Flamels, both hale and hearty, over 375 years old and living in the Indies. Half a century later, the Flamels turned up at the Paris Opéra. 'This last titbit is beautifully reprised in Harry Potter, where Rowling describes Nicolas as an opera-lover and his current age as 665 (a birthday he would have celebrated in 1995 or 1996),' says Principe.

Even though it is doubtful that he performed any alchemy, let alone found the stone, Flamel's work was influential and well known to seventeenth-century alchemists such as Robert Boyle and Sir Isaac Newton. Newton possessed a copy of Flamel's writings and wrote a seven-page summary, entitled 'The Hyeroglyphical figures of Nicolas Flamel explained, anno 1399' as part of his struggle to uncover the true ancient alchemy behind the corrupted versions that had come down to him.

The effort to find the stone may not have seemed so far-fetched at the time, which lies at the watershed between science and magic. There was a widespread belief that metals were made up of a mixture of elementary principles, an idea that had been around since the ancient Greeks dominated science and philosophy. Empedocles, and later Aristotle, developed the theory that all things are composed of four elements: air, earth, water and fire. Thus if an alchemist could find a way to alter the mix, it seemed reasonable to expect that one metal could be changed into another.

In the early modern era, alchemists tended to adopt different elementary principles, comments Principe. Like any self-respecting protoscientist, they had noticed that Aristotle's recipe did not seem to fit what they saw in the laboratory. A widespread belief that all metals were composed of only two elementary principles – sulphur and mercury – in different proportions and degrees of purity emerged in the Islamic Middle Ages, around the ninth century, before being adopted in Europe. By 'sulphur' and 'mercury', however, they did not mean the actual elements but their properties; 'sulphur' was generally regarded as the principle of combustion and also of colour, and was said to be present on account of the fact that most metals are changed into earthy substances with the aid of fire; and to 'mercury,' the metallic principle, was attributed such properties as fusibility, malleability and lustre. In a general way, one can understand how it could have been thought that by combining the yellow of sulphur with metallic mercury, a yellow metal would result. With the correct recipe you could mint gold.

Often, as Flamel's story illustrates, two forms of the philosopher's stone were distinguished, or perhaps two degrees of perfection; that for transmuting the 'imperfect' metals into silver was said to be white, while the stone or 'powder of projection' for gold was said to be of a red colour. In the first Harry Potter book, Voldemort hunts a stone that is as red as blood.

The elixir of life was generally described as a solution of the stone in spirits of wine that would restore the flower of youth. How does it work? Easy. 'The philosopher's stone,' says Paracelsus – a Falstaffian character and chemical pioneer, also known as Theophrastus Philippus Aureolus Bombastus von Hohenheim (1493–1541) – 'purges the whole body of man, and cleanses it from all impurities by the introduction of new and more youthful forces which it joins to the nature of man.'

The alchemical mindset

The heyday of European alchemy was in the latter half of the sixteenth century, when Prague was the 'metropolis of alchemy' and the Holy Roman emperors Maximilian II (reigned 1564–76) and Rudolf II (reigned 1576–1612) sponsored leading alchemists. But there was also official ambivalence about the subject: alchemy posed a threat to the control of precious metal, and The Art was often outlawed. Moreover, many alchemists turned out to be conmen, tricksters and frauds.

Emperor Rudolf II incarcerated a number of charlatans, notably Edward Kelley, whom we encountered earlier. In 1610, growing scepticism about The Art was reflected in Ben Jonson's play *The Alchemist*, where Subtle and his assistant Face attempt to cheat Sir Epicure Mammon to invest in their non-existent gold-manufacturing process.

Even so, eminent scientific figures remained under the sway of sirens of the past whose beliefs in the transmutation of base metals into precious ones still had influence. Although Isaac Newton's laws became the cornerstones of physical science, his alchemical activity, notably his vain search for methods of transmutation, generally remained hidden from the broader public.

Recently, historians and scientists re-examining alchemy have begun to discern genuine chemical insight in its methods, as alchemists struggled not only to understand chemistry but to express what they had found in a systematic way. They used a different language and lacked understanding of what today are seen as basic principles, notably how elements are permanent (which made the idea of transmutation more plausible).

Through careful study of old alchemical formulas, pictures and codes – and by attempting to re-create alchemical experiments – Lawrence Principe and others have begun to decipher some of the enigmas formerly dismissed as mystical nonsense.

The work suggests that some of the leading alchemists helped to lay the foundations of true scientific chemistry.

One illustration, published more than four centuries ago, shows a wolf devouring a king and then mounting a flaming pyre that in turn consumes the wolf to restore the king. Historians now think that this refers to the recipe for refining pure gold from its alloys because one of the traditional alchemical symbols of gold is a king. Experiments by Principe back this view.

The wolf symbolised the mineral stibnite – antimony sulphide – which, in molten form, 'voraciously' dissolves impurities from a molten alloy containing gold. Since antimony sulphide converts "base' metals into sulphide scums that can be skimmed away from the melted gold, the precious metal can be refined from an alloy, such as a coin consisting of a mixture of gold with copper or silver.

Principe has investigated the alchemical leanings of one of the most significant figures to straddle magic and science, Robert Boyle, who lived from 1627 to 1691 and was one of the founders of modern chemistry. Boyle studied the fundamental role of air in respiration and various chemical reactions, and also demonstrated that the volume and pressure of a gas are inversely proportional.

Principe has reinterpreted Boyle's most famous work, *The Sceptical Chymist* (1661), to show that it is not a broad-brush criticism of alchemists, as had been traditionally thought, but an attack on those alchemists who took an 'unphilosophical' approach, that is, those who only cared about making saleable chemical products, such as pharmaceuticals, without giving any thought to how theory might improve their practice. In his book, *The Aspiring Adept*, Principe shows that Boyle was enthusiastic about alchemy in his *Dialogue on the Transmutation and Melioration of Metals*.

Alchemists were struggling not only to categorise their discoveries but to keep them secret from their rivals, which was necessary in the days before patent protection. In the Boyle papers held by the Royal Society in London, Principe examined scraps of paper and discovered that they were partial keys to a code. The papers list the Latin names of chemical substances next to nonsense words that Boyle invented to describe them. When the code is used to decipher Boyle's text describing a chemical reaction, the reaction now makes sense. For example, wherever Boyle scribbled 'ormunt' in a recipe he meant nitre, what we call potassium nitrate. (To complicate matters, Boyle used other codes: for example, replacing words with their Greek or Hebrew equivalents, like 'cassiteros' for tin.)

The kind of intellectual property that Boyle was trying to protect was a solvent called *menstruum peracutum*. He used it to dissolve gold, and when the white residue that resulted was mixed with melted borax, a tiny amount of pure silver was created. His excitement at having transmuted gold into silver must have encouraged him that the quest for the philosopher's stone was achievable. Alas, experiments by Principe suggest otherwise. He followed Boyle's recipe for *menstruum peracutum* – in modern parlance, antimony trichloride and nitric acid – and concluded that Boyle in fact had made metallic antimony, not silver, which is hardly surprising given the ingredients.

Boyle was also persuaded in part that the philosopher's stone was real because others claimed to have witnessed transmutation. One was Wenzel Seyler, a monk from Bohemia who transformed an allegedly silver medallion of Emperor Leopold I and dated 1677 into pure gold when dipped into a 'tincture'. A chemical analysis performed in 1932 revealed that the medal was actually an alloy of gold, silver and copper, so that when dipped into diluted nitric acid, all its component metals except for the gold dissolved to leave a pure gold surface.

Boyle not only believed in the idea of the philosopher's

stone but also thought it would enable him to communicate with angels and spirits, reflecting a spiritual dimension typical of what Principe calls 'supernatural alchemy' developed by one peculiar seventeenth-century school of thought in England. Boyle was pious, and it could be that he saw alchemy as a defence against the growing tide of atheism that tormented him. His belief also seemed to echo elements of the experience of one very prominent 'wizard' who came before him, John Dee, whom we encountered earlier.

Atomic alchemists

Today, the quest to turn one element into another with the philosopher's stone does not sound so far-fetched. Converting lead into gold requires nothing less than the conversion of one chemical element into another. That, in turn, must involve changes to its atoms, where an atom is the smallest unit that bears the chemical characteristics of an element, whether gold or hydrogen.

Each atom consists of a positively charged nucleus – containing neutrons, which have no charge, and positively charged protons – orbited by a mist of negative charge, made up of one or more negatively charged electrons. Atoms are mostly empty space: the nucleus, where most of their mass resides, is 100,000 times smaller than the overall atom.

The nucleus is central to the identity of an element and each one is defined by the number of protons in its nucleus, which is reflected in the element's atomic number. Plutonium, for example, has an atomic number of 94 while gold is 79 and hydrogen one.

Conventional chemistry alters only the distribution of electrons around atoms. To carry the magic of the alchemists, we need to turn to physicists, who have developed various methods to tinker with the nucleus and hence the atomic number of an element. The first artificial transmutation was carried out in 1919 by the New Zealand-born British physicist Ernest Rutherford, who turned nitrogen (atomic number 7) into oxygen (atomic number 8) by bombarding nitrogen with helium nuclei from a radium source. The fast-moving particles penetrated the nitrogen atom and left an extra proton in the nucleus, transforming it into an oxygen atom.

However, only a very few nuclear transmutations could be produced by natural projectiles from radioactive substances. To make nuclear transmutations on a larger scale, and generate new insights into the structure of atomic nuclei, a more powerful stream of projectiles was needed. Accordingly, by the end of the 1920s, there were studies of how to accelerate charged particles to higher energies.

In 1951 the Nobel Prize in physics went to Sir John Cockcroft, director of the Atomic Energy Research Establishment at Harwell, and Ernest Walton of Dublin University for work conducted two decades earlier in which they accelerated hydrogen nuclei towards a lithium target to make helium, marking the first nuclear transmutation produced by means entirely under human control. In subsequent work, Cockcroft and Walton investigated the transmutations of many other atomic nuclei, and the field began to use various types of particle accelerators for the job.

Probably the first kind of transmutation into gold (atomic number 79) was made early in the development of nuclear reactors when platinum (atomic number 78) was irradiated with neutrons. There are many other ways to carry out this feat. Nuclear alchemist Peter Armbruster of the Gesellschaft für Schwerionenforschung, Darmstadt, reported in 1980 that he had amalgamated copper (atomic number 29) with tin (atomic number 50) and observed gold. Bashing particles out of a nucleus can also have alchemical consequences, so that it is possible to turn lead (atomic number 82) into gold (atomic number 79) by removing three protons from the nucleus of each lead atom.

Like so many feats of Muggle science, nature got there first. Nuclear alchemy dates back to the dawn of time. The Big Bang of creation some 15 billion years ago made only hydrogen and then helium, with trace amounts of lithium and beryllium, the lightest and simplest elements, with atomic numbers 3 and 4 respectively. Astrophysicists have long believed that the heavier elements were created in violent nuclear reactions at the centre of stars.

A star's energy comes from the fusion – combination – of light elements into heavier elements, in a process called nucleosynthesis. This astronomical alchemy requires high-speed collisions which can only be achieved with very high temperatures. In the 1950s, Fred Hoyle, William Fowler, and Geoffrey and Margaret Burbidge – and, in parallel work, Alistair Cameron – quantitatively explained almost the entire Periodic Table as the outcome of nuclear fusion in stars and supernovae, when stars explode.

Within these cosmic cauldrons the minimum temperature required for the fusion of hydrogen is five million degrees Celsius. Our sun is burning, or fusing, hydrogen to helium, a process that occurs during most of a star's lifetime. After the hydrogen in the star's core is exhausted, it can burn helium to form progressively heavier elements – carbon and oxygen and so on – until iron and nickel are formed. Elements with more protons in their nuclei require still more energy. Most of the heavy elements, from oxygen up through iron, are thought to be produced in stars that contain at least ten times as much matter as our sun.

Supernova explosions result when the cores of massive stars have exhausted their fuel supplies and burned everything into iron and nickel. Elements with mass heavier than that of nickel are thought to be formed during these explosions. Thanks to the transmutations of cosmic alchemy we are all, in effect, made of stardust. Now that is magic.

The elixir of life

Like their forebears in ancient China, today's alchemists still think the secret of longevity is within their grasp, thanks to a research effort that has severed any link with the stone and its promise of nuclear transformation. Age at death is at least partly under genetic control, and the race is on now among Muggle scientists across the planet to identify the relevant genes and molecular mechanisms so that they can be manipulated to create an elixir of life, or perhaps some kind of anti-ageing pill.

The very first life-extension genetic mutation, in a gene called *age-1*, was found in the tiny nematode worm, *Caenorhabditis elegans*, one of the most popular experimental subjects for this kind of work. Since the gene was discovered by Michael Klass at the University of Colorado in Boulder, others have been uncovered. Studies by Cynthia Kenyon's 'worm group' at the University of California, San Francisco, have shown that *age-1* and *daf-2* mutations act in the same life-span pathway, and suggest that how fast worms – and humans – age is related to how we burn the calories we eat. Reducing the number of calories consumed without inducing malnutrition is a proven method in a range of species for extending life span and postponing ageing. Some have suggested that average life span could be extended by twenty years if people cut their calorie intake by 30 per cent. No surprise, then, that the equivalent of the *daf-2* gene in humans produces the docking point on cells that responds to insulin and insulin-like growth factor, which play a role in regulating food metabolism. 'It may be that one portion of this pathway is involved in the control of ageing, so the challenge is to figure out which portion that is,' says Kenyon. 'These studies show that hormones control ageing,' she adds, explaining that these hormone pathways probably evolved to allow animals to postpone reproduction when conditions were harsh and offspring likely to perish.

Smell and taste may influence life span, too. Kenyon's team found that lessening the worm's ability to perceive its surroundings increases its life span by more than a third. Instead of living for two-plus weeks, the animals live three or four weeks – the equivalent of a jump of from, say, 90 to 130 years in humans – and remain youthful.

This does not mean that we should restrict our diet – say to pork sausages alone – to stay young. What it does suggest is that chemical signals from the environment, perhaps chemical signals called pheromones or the smell or taste of something in the nematodes' food, influence their rate of ageing, probably by acting on a hormone signalling system. The finding fits with an earlier one that demonstrated how signals from the reproductive system influence the life span of the worm.

There has always been a link between sex and death because the body's job seems to be to keep an individual alive long enough to pass on his or her genes: sperm and eggs are designed to perpetuate the family line, and the things they are attached to – bodies – are merely disposable vehicles to carry this goal out. Kenyon found that signals from germ cells – the sperm and egg in humans – shorten life span, while signals from the somatic gonads (reproductive tissue that surrounds the germ cells) lengthen life span, sending equal-but-opposite signals. If the cells that give rise to the germ line are removed, the animals remain youthful longer, and live much longer than normal. (Kenyon adds: the longevity is not linked to a lack of children, since destroying both germ and somatic cells leaves life span unchanged.)

Another key theme of anti-ageing research is to uncover the role of tissue-damaging chemicals called free radicals, which cause 'oxidative stress'. The hope is that by resisting stress and repairing damage, ageing can be slowed. Fruit flies usually live for about forty-five days, but if a single gene is mutated the insects can shrug off stresses such as heat, starvation and radical-forming herbicides and live 35 per cent longer. Seymour Benzer of Caltech called it the 'Methuselah gene', after the Old Testament character who lived for 969 years. In other work, genetically engineered fruit flies, which produce extra antioxidants to protect against free radicals, were found to live about seventy-five days and some have lived as long as ninety-five days.

Other worm studies have come to similar conclusions. One shows that by mimicking two of the body's natural defences against toxic oxygen radicals, drugs called synthetic catalytic scavengers can boost the worms' life span by 50–100 per cent. 'We were amazed by what we were seeing down the microscope,' says Gordon Lithgow from the Buck Institute, California, one of the team. 'As the untreated worms began to die, the drug-treated worms were swimming around, full of life.' The discovery that a drug can prolong life span 'increases our confidence that ageing is a solvable biological problem', asserts Lithgow.

The same seems to work for mammals, and presumably Muggles too. Methuselah mice that are resistant to radicals – and lived one third longer – were created by deleting a gene called p66shc. Elderly rats given two dietary supplements – acetyl-L-carnitine and an antioxidant, alpha-lipoic acid – had more pep, better memory and 'got up and did the Macarena', says Bruce Ames of the University of California, Berkeley.

Another focus of efforts to find the elixir of life is a drug that can manipulate structures called telomeres on the ends of chromosomes, the packages of genes in cells. Rather like the plastic bits at the end of shoelaces, telomeres stop chromosomes from 'fraying'. Telomeres shorten with repeated cell division until, when they have dwindled to a certain point, the cell can no longer divide. A handful of human cells – germ cells, cancer cells and embryonic stem cells – use an enzyme called telomerase to circumvent that shortening process, and thus to also circumvent the ageing process and achieve a cellular version of immortality.

While drugs offer one way to tinker with telomerase, cloning offers another. Telomeres from cows cloned by Advanced Cell Technology of Worcester, Massachusetts, were found to be longer than those from normal cows of the same age and, in most cases, even longer than those from newborn calves. Far from being prematurely aged, cells from the cow clones appear to have recaptured and even prolonged their youth, perhaps lengthening their life span beyond that expected for their chronological age.

The only problem with the focus on telomeres, however, is that even the cells of the oldest old still have quite a few more cell divisions left to go, suggesting that there is more to the anti-ageing story than cell division alone. Another worry is that telomerase is activated in 90 per cent of all cancers, an uncontrolled growth in which cells continue to divide at a high rate.

Many human genes have also been linked with ageing. The first was responsible for a rare disease called Werner's syndrome. People affected by this disease appear to age in adolescence. They become grey, develop wrinkled skin, then lose their hair, usually dying before the age of fifty. American researchers found that they carried a faulty version of a gene for a type of enzyme known as a helicase. The enzyme splits apart or unwinds the two strands of the DNA double helix, a process that must take place in healthy dividing cells if they are to pass on their genetic material, in the form of chromosomes, to daughter cells. If a broken helicase accelerates ageing, it could be possible that, by mending it, sufferers would live longer.

Diverse age-related diseases –- including cancer, atherosclerosis, arthritis and Alzheimer's disease – may have a common link in a gene called p21, says Igor Roninson at the University of Illinois at Chicago. The p21 gene serves as a brake that stops cells from growing when they are damaged by toxins or radiation, giving them time for repair. During normal ageing, the p21 gene also stops cells from dividing as the cells grow old. But the team found that turning on this single gene brought about changes in numerous others: it damps down more than forty genes known to be involved in DNA replication and cell division, arresting cell growth. At the same time, p21 increased the activity of about fifty other genes, some linked to cancer or to age-related diseases. As a result, cells stopped growing, became flat and granular in appearance and started making enzymes typically produced by elderly cells.

As each new genetic insight suggests another way to arrest ageing, other research has focused on how to grow new cells and tissue to keep a body going long into old age. As discussed in Chapter 6, one ideal source is to use stem cells from the early embryo, when it consists of around 100 cells. The adult body also contains stem cells, such as those in bone marrow that give rise to blood cells, though their potential seems more limited.

The role of these cells in ageing was underlined by a study that found, paradoxically, that too much of a cancer-preventing protein leads to premature ageing in mice, a discovery that suggests that the body may have to strike a balance between preventing cancer and succumbing to old age (although this 'trade off' theory is disputed by others). Lawrence Donehower of Baylor College of Medicine, Houston, and colleagues created mutant mice in which an overactive form of the p53 protein was made by accident. This should have been good news: p53 is one of the cells' key lines of defence, halting cell division, repairing DNA damage and triggering cell death if cancer develops. As expected, the mice developed fewer tumours than their normal counterparts, but they did not live longer. Their average life span was 96 weeks, compared with 118 weeks for normal mice – a reduction of nearly 20 per cent. The mutants lost weight and muscle, suffered hunched backs and brittle bones, and their wounds took longer to heal. The team suspects the excess p53 stunts the division of stem cells that normally replenish tissues such as skin and bone in adults, causing premature ageing.

The embryonic stem cell has vast potential – the genetic recipe and the biological know-how – to become any cell, any tissue, and any organ in the body. With the right biochemical signals, it could be coaxed into forming muscle, which could replace tissue damaged by a heart attack. Or into islet cells to treat diabetes. Or into retinal cells, which could be used to restore fading vision. Or into endothelial vessels to replace those clogged by deposits. Or into brain cells, which could be used to treat Parkinson's disease. The stem cell is so versatile and potent that it could conceivably be used to grow an unlimited supply of body parts – whether sinew and bone, or blood and brain – to repair the tatters of old age.

The quest for the elixir of life continues. Although we are far from being able to achieve the fictional life span of the Flamels, progress is being made on a range of fronts. Voldemort once told the Death Eaters that his goal is to conquer death. No doubt, he would approve of this particular Muggle magic.

CHAPTER 14

Belief, superstition and magic

Magical inventions are so frequent and so widespread that we should ask ourselves if we are not confronted with a permanent and universal form of thought.

Claude Lévi-Strauss

Much of the magic in the Harry Potter books may seem absurd, whether it's lawnmowers with unnatural powers, cursed hats that make ears shrivel or a Latin spell that turns pineapples into tap dancers. The feat of turning teacups into rats also sounds quite fantastic, as does transforming rabbits into slippers. However, it is too glib to dismiss all these abilities as daft or silly (though most undoubtedly are), just as it is facile to reject all ancient myth, folklore and legend as the products of a primitive mind, as beliefs held by simple people that had no possible basis in reality.

Throughout this book we have seen evidence that the worldview of a scientist can seem as peculiar as the worldview of a shaman, witch or medicine man. Scientific papers often seem as obscure as anything in the occult literature. Many of them deal with the intangible, the abstract, the theoretical and invisible to the point that you could be forgiven for wondering if they discuss anything real at all.

In Chapter 8, I discussed the origins of belief and superstition. Belief in something – anything – has been a common element of every society and every culture since the dawn of humanity. This simple fact raises many questions. When did this facility emerge? How do beliefs spread? Why do they persist?

There is evidence that belief in anything, no matter how bizarre or unlikely, is good for us. In that sense, irrational beliefs can be considered quite rational, as rational as a belief in science. That raises more questions. If belief is indeed is good for us, do we learn how to believe or are we born with a propensity to believe? Does this ubiquitous desire to have faith in something have a genetic component?

The universality of belief also raises questions about the nature of science and how it differs from faith. The quest for the fundamental laws of the universe is pursued with an almost religious zeal. Some even believe that it will give some insights into the 'mind of God'.

However, we will discover that there are fundamental reasons why we can't proceed from those laws to explain everything, as some claim. Superstring theory, or whatever brand of 'final theory' (or 'theory of everything') is currently in vogue, will never have the capacity to predict the migratory habits of the gnu the appetite for Harry Potter. Indeed, the very foundation of science, mathematics, contains randomness, a finding that should trouble any triumphalist boffin who expects to tie the ultimate secrets of the universe into a neat bundle of theory, with no dangling loose ends. Is science then a matter of faith, a mathematical voodoo, a theoretical fetish, a kind of religion based on research?

Quantum sorcery

If science has taught us one thing it is that common sense is an unreliable guide to the way nature works. The world can look downright peculiar through the eyes of a scientist. The 'common sense' that is hard-wired into our brains has evolved to deal with problems that were common on the African savanna where our ancestors started out. What was useful making sense of the immediate surroundings then, from flying spears to charging lions, is not necessarily useful for providing an objective understanding of today's wired world.

The prototypical example is the belief that the sun goes around the earth: an 'obvious' induction backed by the experience of billions of people every day – but of course false. If a sharpshooter drops a bullet and, from the same height, fires a bullet horizontally, the projectiles will hit the ground at the same time, neglecting air resistance (the impetus provided by gunpowder contributes nothing to the bullet's vertical motion). But most people would take the commonsense view: of course, the fired bullet will arrive first. Nor are our brains very good at handling probabilities. Spin a coin six times: many will think that alternating tails and heads, after a first head, is more likely than six heads in a row. And they would be wrong.

The departure of common sense from science has much to do with mathematics. Consider the most successful branch of science, quantum theory, which provides a mathematical worldview that when translated into everyday language is headache-inducing and defies common sense. In our everyday world, we expect effects to have causes. Yet quantum mechanics admits unpredictable hops – 'quantum jumps'.

In our everyday world, there seems to be no limit to how accurately we can measure the properties of an object like a broomstick, such as its weight or dimensions. Not so in quantum theory. The uncertainty principle, enunciated by Werner Heisenberg in 1927, states that measurements of certain pairs of quantities, such as position and momentum, can be made only to a certain degree of precision, and no further.

According to this unsettling and strongly counter-intuitive theory, all physical objects are intrinsically ghostly, as spectral as the spirits that haunt Hogwarts. They exist in a twilight state – a 'superposition' (combination) – of all possibilities of position and velocity. Only when a measurement is made on an object do we gain information about specific values of its observable properties.

Particles of matter are waves of energy, and waves are particles. Each can appear as one or the other depending on what sort of measurement is being performed. Stranger still, a particle moving between two points in space travels along all possible paths between them simultaneously. Indeed, the behaviour of particles that are at opposite ends of the universe cannot be described separately by quantum lore. Don't feel too intimidated by all this: even the great Albert Einstein, after helping to establish quantum mechanics a century ago, developed a profound distaste for the field.

The reason for assaulting your common sense this way is to show how science sets out with a quite reasonable urge to understand the way the world works, based on the quite reasonable view that things like tables are what they seem, and a quite reasonable assumption that we perceive them directly. However, as a result of scientific inquiry, this commonsense view has given way to the quite unreasonable picture of tables consisting of invisible particles that we can only perceive indirectly, by seeing other particles (light particles called photons) colliding with our retinas. As Lord Haldane, the British scientist, once said, 'My own suspicion is that the universe is not only queerer than we suppose but queerer than we *can* suppose.'

Because common sense is flawed, the wildest magical beliefs really do not seem so different from those cherished by scientists. Is the cosmological theory of creation, which says the universe was born billions of years ago in the Big Bang, really more believable than the idea of a wizard materialising in a fireplace? The Judeo-Christian creation story begins with a void, which is nowhere near as strange as modern cosmological theory, which begins with absolutely nothing, not even space itself. No wonder that a belief in magical abilities and supernatural powers has been prevalent across all cultures from the darkest days of prehistory.

The anthropology of belief

The Dursleys and no doubt other inhabitants of Little Whinging in Surrey would be depressed by what I am about to discuss, given their deep hatred of all wizardly things. But anyone with an inquiring mind should be struck by how magic is so pervasive, a point that is no doubt made by Bathilda Bagshot in her legendary tome, *A History of Magic*.

The first attempt to carry out an extensive study of magical belief was published in 1890 by the Scottish classicist Sir James Frazer. *The Golden Bough* marked a milestone in anthropology, its dozen volumes revealing the variety and extent of magical beliefs while discerning their common features, such as positive magic, or sorcery, which is summoned to bring about desirable ends, and negative magic, or taboo, which aims to avoid unwanted consequences.

Frazer identified homeopathic magic, the idea that 'like produces like' – for instance, when an American Ojibwa hits a doll modelled on an enemy to harm that foe – and contagious magic, which is based on how there is a lasting connection between things that were once in contact. In parts of Germany it was once thought you could make someone lame by driving a nail through his footprint.

In *The Elementary Forms of Religious Life* (1912), the French sociologist Émile Durkheim placed superstition in a social context, taking into account the cultural status of science and technology. The use of the term 'supernatural' assumes that a society understands the natural order, the laws of nature that have been elaborated by science. But, of course, not all cultures distinguish between the two, so Durkheim offered another distinction, that of the sacred – the stuff of magic and religion – and the profane, the stuff of the secular and science.

Bronislaw Malinowski, whose interest in anthropology was sparked by reading *The Golden Bough*, took the seemingly obvious step of living among a magical culture – in his case the Trobriand islanders in the southwest Pacific during the First World War – rather than relying on second-hand accounts of their culture from missionaries and ethnographers. During his 'participant observation', Malinowski noted how the Trobrianders would perform magical rites over their gardens to ensure horticultural success even though they had rudimentary scientific knowledge about soil types, seedling selection and planting methods. They knew that, even if they did everything just right, their crops could still fail as a result of nature's capriciousness. Magic was used to head off bush pigs, locusts, drought and other misfortunes. 'Magic is akin to science in that it always has a definite aim intimately associated with human instincts, needs and pursuits,' said Malinowski. 'Thus both magic and science show certain similarities, and, with Sir James Frazer, we can appropriately call magic a pseudo-science. The theories of knowledge are dictated by logic, those of magic by the association of ideas under the influence of desire.'

Malinowski offered a pragmatic view of magic and superstition as a way to cope with the anxieties of life. 'Both magic and religion arise and function in situations of emotional stress: crises of life, lacunae in important pursuits, death and initiation into tribal mysteries, unhappy love and unsatisfied hate. Both magic and religion open up escapes from such situations.'

In 1954, Malinowski summed up the relationship of magic, religion and science as follows:

> Science, primitive knowledge, bestows on man an immense biological advantage, setting him far above all the rest of creation . . . religious faith establishes, fixes and enhances all valuable mental attitudes, such as reverence for tradition, harmony with environment, courage and confidence in the struggle with difficulties and at the prospect of death . . . The function of magic is to ritualise man's optimism, to enhance his faith in the victory of hope over fear. Magic expresses the greater value for man of confidence over doubt, of steadfastness over vacillation, of optimism over pessimism.

Sir Keith Thomas, a great British historian, examined magical beliefs and practices in England between 1500 and 1700. In *Religion and the Decline of Magic*, he highlighted contributing factors in society that led to magic's demise. The priests of the medieval church won converts by incorporating pagan supernaturalism, blurring the distinction between magic and religion. There was a widespread belief that consecrated objects, even the soil of the churchyard, had magical powers. Holy water was splashed around to ward off evil spirits.

The waning influence of the supernatural and witchcraft in the eighteenth century resulted from a variety of influences, including the developments that made life more secure, an increase in literacy, and advances in science and technology. Rather than relying on mystical incantations and so on, people became more self-reliant and put faith in their own initiative. Even so, magic will never become extinct. Given the ever-present worries and concerns of daily life it is telling that Thomas remarked: 'If magic is to be defined as the employment of ineffective techniques to allay anxiety when effective ones are not available, then we must recognise that no society will ever be free from it.'

We are not free from it today. Psychologist Stuart Vyse, in *Believing in Magic*, cites the growth in the number of books listed under 'occult and psychic' from 131 in 1965 to 2,858 in 1994, the rise of the bland one-size-fits-all predictions of newspaper astrologers, and the evidence of Gallup polls which revealed, for example, that 72 per cent of Americans believed in angels in 1994. 'As a group, contemporary Americans may be less superstitious than Britons of the sixteenth and seventeenth centuries, but science and reason have yet to defeat the forces of the paranormal. Indeed, many recent victories have been on the side of superstition.'

Vyse wrote those words in 1996, and I asked him if he had revised his opinion since then. 'Alas, this picture has not changed. My impression is that paranormalism is still a growth industry. Here in the US there has been increased interest in Nostradamus since the attack of September 11 (of course, he was said to have predicted it, according to an e-mail widely circulated on the internet that was later discovered to be a hoax, or at least a fabrication), and psychics who claim to be able to speak to the dead are enjoying great popularity. One has a popular television show.'

The Harry Potters still enormously outnumber the likes of Petunia and Vernon Dursley in society, who just don't hold with such nonsense, perhaps reflecting how modern life seems more out of control than ever. Most people do indeed believe in magic of one sort or another, whether the thespian who shouts 'Break a leg' at a colleague, the student who always wears the same outfit for exams, the blushing bride who crosses her fingers for good luck, or those who jump with joy when they find a four-leaf clover. Why is our belief in magic so deeply ingrained? Indeed, why do we believe in anything at all?

The origins of belief

The first place to look for belief and the stirrings of an ability to link cause and effect is among the monkeys and apes, which provide a good benchmark to judge our own behaviour. If an ape can link a strong breeze with the movement of a tree, a meal of bitter-tasting pith with the subsequent disappearance of gut ache, or success in obtaining a tasty morsel with the design of a tool, then it must have some kind of belief.

Apes can learn many complex tasks. Chimpanzees can even use simple tools to get ants from cracks and smash nuts open with pounding tools and anvils. Can they link cause and effect? According to Frans de Waal of the Yerkes Primate Center in Atlanta, they can to some extent. He points out, for example, that if you travelled through the canopy of a forest with an ape, you would notice that it had a better understanding of where to go, and which branches to choose, than *Homo sapiens*.

The ability to link cause and effect varies among primates. One study by Elisabetta Visalberghi of the Consiglio Nazionale delle Ricerche, Rome, investigated tool use by human infants, capuchin monkeys and chimpanzees to shed light on causality. She studied how each species used a stick to push a morsel of food out of a tube. When her subjects became proficient at using the stick, she changed the task so that the middle of the tube contained a trap. Depending on the side in which the primate inserted the tool, the animal could either push the reward into the trap or out of the tube.

Understanding causality requires the primate to understand more than just that two events are associated with each other in space and time. There must also be an idea of a 'mediating force' that binds the two events which may be used to predict or control those events – a physical force that pushes the food to the trap and makes it fall into the trap, or a 'psychological force' such as an intention. In other words, causality requires a mental model of how the world works rather than a vast database of examples of where effect X followed cause Y.

In the tube task, the success rate matched that expected by chance for all capuchins except one, who discovered that it would always get food by inserting the stick from the end of the tube farthest away from the reward. Needless to say, explains Visalberghi, 'This magic rule turned out to be a disaster as soon as the experimenter changed the lengths of the tube arms to control for it.' The chimpanzees did better than chance, suggesting they had a better understanding of cause and effect. They required many trials to solve the task above chance level, but they eventually came to a more flexible understanding. Children under the age of three could not figure out a successful strategy, unlike those aged over three.

She concluded that the capuchins have a weak causal understanding of the way tools work; they have no understanding of the physical forces but learn to associate their actions with results. Apes, on the other hand, do show some possible signs of understanding the causal relations involved in tool use – although not to the extent of human children. This work suggests that the concept of causality is most fully developed in human beings.

In the social domain, an understanding of causality has other consequences. Humans can 'mind read', that is, that they can work out the intentions of others. A theory of mind is a short step from imputing a mind that exists *independently* of a body, such as a soul, devil or god, according to the evolutionary psychologist Steven Pinker. In this way, the instinct to link cause and effect could have sown the seeds of magical thinking, where a spoiled harvest could be linked to a malevolent spirit, for example. This willingness to animate the inanimate can be seen clearly in children. As the Swiss psychologist Jean Piaget once pointed out, children bestow inanimate objects with a free will of their own, lending a magical aspect to their thought processes.

Human action was the only causal agent that ancient people

were really certain existed, so it was natural to assign the causes of inexplicable events to a human-like agent, argues anthropologist Stewart Guthrie of Fordham University. This tendency to anthropomorphise stems from a deep-seated perceptual strategy: in the face of pervasive (if mostly unconscious) uncertainty about what we perceive, we bet on the most meaningful interpretation we can. If we are in the woods and see a dark shape that might be either a member of an enemy tribe or a boulder, for example, it is a good strategy to think it is a enemy. If we are mistaken, we lose little, and if we are right, we gain much. So, in scanning the world we always look for what most concerns us – living things, and especially, humans.

'It is not merely that we fear other humans more than other creatures, though that is an important factor. Rather, it is also that we cherish and value them more as well, for complex reasons including practical ones,' says Guthrie. 'Is that rustle in the bushes a person who will attack me? Is that glint in the sky a rescue aeroplane coming to save me from this desert island? Is that distant bit of maroon the jacket of my missing child?' Humans and human-like models, whether spirits or gods, offered our ancestors the most far-reaching interpretations and explanations of the world.

The birth of magical thinking

When was belief born in human beings? While genetic and fossil evidence suggests that humans were anatomically modern in Africa by over 100,000 years ago, scholars have argued long and hard over whether human behaviour, such as belief in magic and toolmaking, developed in tandem with physique. Some insist that behaviour typical of modern humans arose relatively late and rapidly, some 40,000 to 50,000 years ago, while others believe that it evolved earlier and more gradually.

Now, however, one might even hazard a guess as to when magical thinking became commonplace, in the light of a recent discovery at Blombos Cave, a site on the southern Cape shore of the Indian Ocean 180 miles east of Cape Town. Earlier work at the site revealed evidence of fishing, manufacture of finely crafted bone tools, sophisticated manufacture of bifacially flaked stone tools, and use of ochre, possibly for body decoration.

Then archaeologists found two pieces of ochre from the Middle Stone Age layers. The pieces of the red ochre, measuring two and three inches long, had been scraped and ground smooth to create flat surfaces. They were then marked with cross hatches and lines to create a consistent complex geometric motif. So far, so unremarkable. Except for their great age.

Dating of the sand grains that lie above the ochre and burnt stone found in the same layer as the works of 'art' revealed that they were engraved more than 70,000 years ago, and were twice as old as the previous record holder. 'The presence of the engraved objects and other evidence (bone tools, bifacial points and so on) signifies the cognitive abilities and capacity for abstract thought are in line with what we would expect of modern human behaviour,' says Christopher Henshilwood of the Iziko-South African Museum.

African people, from whom we are all descended, were modern in their behaviour long before they arrived in Europe as Cro-Magnons and replaced the Neanderthals. 'I think the cognitive ability for abstract thought and abstract depiction would certainly allow modern humans the facility to extend into a spiritual/magic/religious world,' says Henshilwood. 'I really doubt that these engravings were idle doodles but must have had meaning and significance to the maker and to his/her fellow cavemates. Could we begin to interpolate these finds as the earliest known beginning of religion, magic and so on? It is certainly possible.'

'You can't make a complex tool without a concept of cause and effect,' adds Lewis Wolpert of University College London.

He argues, like Malinowski, that one striking side effect of our well-developed ability to link cause and effect was the emergence of belief. Belief evolved because it allowed our ancestors to cope with nature's deadly vicissitudes. Of all the important events for which ancient people could find no clear cause, death was the common denominator. Life was tough and many children died at a young age. Volcanoes, storms, lightning, floods, climate change and other natural disasters abounded.

The ancients must have felt uncomfortable about their inability to control or understand such things, as indeed we do today. As a consequence, they began to construct, as it were, false knowledge, says Wolpert. To deal with an uncertain and unpredictable world, magic, a belief in the supernatural capacity of humans to manipulate nature, and religion, a belief in divine beings that act independently of the human race, emerged.

The primary aim of our first faltering attempts to model the way the world works was not accuracy, the sole motivation of scientists, but rather the avoidance of paralysing uncertainty. In this way, Wolpert argues that it may have been fear that produced superstition, ritual, magic and gods. Linking events (the death of a relative, the failure of a crop and so on) to causes (some humanlike agency) gave our ancestors two adaptive advantages: uncertainty, and thus anxiety, were removed, and there was now an animate agent that might be appeased in some way.

Much evidence exists that superstition thrives on the human desire for control, which has been shown to reduce stress. There are studies, for example of breast cancer patients, which show that those who believe they are in control fare better than those who do not. One investigation of 'magical thinking' during the 1991 Persian Gulf War by Giora Keinan of the University of Tel Aviv found that people who lived in the more dangerous areas of Israel were more superstitious. There are other examples where a sense of control, even if illusory, can help deal with stress, according to the psychologist Stuart Vyse. 'As Nature abhors a vacuum, so does human nature abhor randomness. We prefer order over chaos, harmony over cacophony, and religion over the prospect of an arbitrary world.'

Superstition is alive and well today because many aspects of life remain out of control, whether they are the questions that come up at an exam or the complications facing a birth. With some exceptions, most superstitions are benign and offer some psychological benefits. Even if real control is not possible, an illusion of control will do. In this sense, the most irrational superstitions are in fact rational. Your eccentric beliefs are good for you.

Multiplying magic

Once a belief is born, it spreads by an unconscious human drive to imitate others, just as we often adopt the same posture as someone we are talking to, or smile back at a grinning gap-toothed toddler. As a result, a belief can invade the human mind like a parasite, according to Richard Dawkins of Oxford University. He likens beliefs, superstitions and so on to a virus of the mind, where the virus contains a program – a set of genes – that says 'Duplicate me.' Adult viruses include chain letters, pyramid selling, urban myths and fashion. Mind viruses of the past include the conviction that the great pestilences were the works of a malevolent devil or a witch or the wages of sin.

The readership of Harry Potter is especially vulnerable to infection by a mind virus. Just think of the popularity of childhood crazes, whether for collecting Chocolate Frog and Pokémon cards, spinning hula hoops or bouncing on pogo sticks. 'This is because children need to be especially receptive. A child has so much to learn. Not just a language but an entire culture has to be transferred into the child's brain. Just as cells and computers are vulnerable to parasitic programmes because they need to be "programme-friendly", so a child's brain is especially at risk.'

Crazes, fashions and Pottermania are the measles and whooping cough of the mind, spreading horizontally, from child to child. Indeed, sometimes parents even help their children to become infected by encouraging activities that are imaginative and use fantasy in play. Perhaps it is these magical experiences of youth that seed adult superstitions.

These are also culture viruses whose pattern of infection is vertical: from parent to offspring, rather like ordinary genes. These range from regional accents to religion. Dawkins famously coined the term 'memes' in his 1976 bestseller *The Selfish Gene*. In the years since, the concept of memes has received a mixed press. Some, like Susan Blackmore of the University of the West of England, argue that imitation is what makes humans special and from this one simple mechanism it is possible to shed light on the evolution of the human brain, the origins of language, altruism and the evolution of the internet. 'Looked at through the new lens of the memes, human beings look quite different.'

Others wonder if memetics are real. Though a fan of both Dawkins and Blackmore, Vyse says the notion of a meme is 'a poorly supported metaphor'. The philosopher Mary Midgley dismisses them as 'mythical entities'. Memes are 'meaningless' according to Stephen Jay Gould, the late evolutionary biologist who was so eminent that he appeared on *The Simpsons*. To reduce magic, with its complex web of artefacts, rituals and customs, and its relationship with society, to simple units – genes and memes – is 'an explanatory imperialism' that represents nothing less than dumbing down, according to archaeologist Timothy Taylor, who cites Oscar Wilde: 'The truth is rarely pure, and never simple.'

At the risk of being drawn into the meme wars, I will only

suggest that memetics is unlikely to be an all-encompassing theory of human nature but, at the very least, it offers a pragmatic way to clarify some ideas about the origins of magic without getting overwhelmed by a fog of wands, rituals, pointed hats and other clutter.

Dawkins points out, for instance, that one reason for the success of many convictions is the accompanying belief that it is virtuous to believe in something precisely because there is no evidence for it. 'The victim of a successful transgenerational mind-virus typically experiences a strange, inner conviction, a deep internal certainty that something must be true, even though there is no evidence for that something. This internal certainty, utterly unshakeable yet wholly free of supporting evidence, is often given the name "faith".'

Dawkins holds back from applying the pejorative virus metaphor to all aspects of culture, all knowledge, all ideas. There is a spectrum from the pure virus at one end, such as urban myths and addictive computer games, to the useful and genuinely desirable programme at the other: 'In the domain of culture, great ideas and great music spread, not because they embody instructions, slavishly carried out but because they are great.' Newton, Darwin, the Brontës, Eliot, Beethoven and Bach produced such great works. And today the same must be said for J.K. Rowling, whose particular mind virus exploits a fascination with magic that dates back many thousands of years.

Belief genes

The idea of genes being linked with beliefs is not so far-fetched, given the profound influence of genetics on the developing brain. There are many suggestions as to why 'belief genes' thrive. The sociobiologist E. O. Wilson points out that 'the predisposition to religious belief is the most complex and powerful force in the human mind' and that religions often help perpetuate their followers' genes by encouraging them to have big families and including prohibitions against potentially contaminated food, incest and other risky activities.

Steven Pinker says that religion can be a mechanism by which some people can take advantage of others. For example, ancestor worship reinforces the power of the old, for they can threaten the young even after their death. Initiations and sacrifices help weed out social parasites. Religion also promises better health, wealth, protection or eternal happiness in an afterlife, and any dogma that promises such miracles can perpetuate itself in a culture.

Although Richard Dawkins attacks religion as an infectious disease of the mind (which certainly seems the case for the suicidal Jonestown and solar temple cults) there is evidence that religion can be beneficial. Those who are more inclined to believe may also have survived better than those who did not have such beliefs. A wide-ranging survey of scientific evidence of the 'faith factor' in disease has been conducted by Mayo Clinic researchers, who concluded that a majority of 350 studies of physical health and 850 studies of mental health have found that religious involvement and spirituality are associated with better health. Overall, 75 per cent report a positive effect, 17 per cent report no effect and 7 per cent report a negative effect.

There are many explanations other than the obvious one of divine intervention. Belief can help people to cope with stress and religious people may be more compliant and less likely to overindulge, or they may be able to draw on a bigger support network (such as a congregation). Wolpert has gone so far as to suggest that religion may be not the opium of the people, but the great placebo of the people (though some scientists remain sceptical about the health benefits of religious activity).

If belief and spiritual factors really do boost an individual's chances of survival, any genes linked with a propensity to believe would survive in future generations. This seems to be the case. The proportion of scientists who believe in god has remained at about 40 per cent during the last eighty years, despite all the extraordinary advances made in research and the supposed decline in religious practices. Thomas Bouchard of the University of Minnesota conducted a study of twins who had been reared apart and concluded that there was 'a modest degree of genetic influence' in two measures of religiousness.

The idea that religion, belief and faith have a genetic component is supported by John Burn, medical director of the Institute of Human Genetics, University of Newcastle, England: 'As a geneticist I can't help noticing that every tribe and every people has religion. If not in official form, they display a capacity for shared beliefs. In our modern secular state, beliefs are more dispersed; our preoccupation with the memorial to Diana, Princess of Wales, and the intensity of Geordie support for Newcastle United [his local football team] might be seen as manifestations of a need to believe.'

The idea that faith is a distinguishing feature of humans has been suggested for as long as the issue has been pondered. It's not difficult to see the potential benefits of such a behavioural trait, says Burn. 'Survival of our species has demanded a capacity to work together, to form societies. A willingness to live, and if necessary die, for a belief is a powerful selective advantage. There is a genetic propensity for us to believe.' Is this perhaps why Muggles across the planet also believe in science?

CHAPTER 15

The magic of science

I have often admired the mystical way of Pythagoras, and the secret magic of numbers.

Sir Thomas Browne

Is science simply another system of beliefs with no more validity than the idea that the earth is flat, that a blood sacrifice is needed to make the sun rise, or that the wisdom of the ancients can be absorbed by eating a dead relative's brains? The answer is no, and the difference comes down to experiments – lots of them. Conventional scientific knowledge is obtained by painstaking slog. Eureka moments and flashes of insight are rare. It can take years to discover something and confirm it to everyone's satisfaction.

This scientific process has, for example, confirmed the bizarre quantum worldview encountered earlier in this book. Yet, as the great Richard Feynman once remarked, 'Nobody understands quantum mechanics.' Most don't really care that it is so out of kilter with everyday experience. The details of quantum theory may seem to be a chapter straight out of *Important Modern Magical Discoveries*, but they do provide a dazzlingly accurate description of the subatomic world, not just of electrons in transistors and the movement of light particles in fibre-optic cables but of chemical and nuclear reactions, and much else besides. One does not have to fret about whether ghostly superpositions and the other troubling features of quantum theory really exist, but only if they provide a good description of the universe. Crucially, quantum thinking has survived decades of testing dialogue between theory and experiment.

By contrast, magical insights seem to be easily obtained, without special knowledge or training. Anyone can be an expert on the paranormal or the supernatural because they place a premium on personal experience. This inclusivity is powerfully comforting and attractive to those who feel intimidated by the mathematical worldview of science.

It is easy to see why people are so intimidated by science. Take Albert Einstein's dream of a single all-encompassing theory of the universe, for example. Today, scientists across the world continue this quest. They want to find a theory of everything that unites all the particles and forces in the cosmos with a few equations that are so concise they could be silk-screened on to a T-shirt.

The quest builds upon one of the first great achievements of science: Newton's unveiling of a theory of gravity in which the laws that governed motions on earth also seemed to rule the heavens. This marked the start of a long journey in which mystery seemed to be gradually being erased. Newton's theory has been overtaken by Einstein's view of gravity, which can handle the more extreme conditions found in the cosmos. Today, the quest for a theory of everything depends on combining theories of the very small (quantum theory) and the very large (general relativity) to give us quantum gravity.

When that dream is realised, will we be able to explain everything? Will the theory extract the mystery of life as surely as a Dementor sucks souls? Will the birth of the theory mark the death of magic? People who are intimidated by science often resent it for the way it supposedly squashes our sense of mystery (a claim eloquently countered by Richard Dawkins in his book *Unweaving the Rainbow*).

We have already seen that superstition thrives on uncertainty, and that is bound to remain however many advances science makes. Perhaps more surprising, there are even mathematical and scientific reasons to believe that magic and mystery will prevail. Those who are leery of science should take heart perhaps from the growing realisation in the past century that, in one sense, science rests on an article of faith, as does a belief in the supernatural. There is something profoundly magical lurking in the logical heart of mathematics, the language of science.

A century ago, it was thought that the power of mathematics was limitless. Mathematicians, notably David Hilbert, had believed that every question could be resolved and shown to be true or false. That dream died on 17 November 1930, when a journal called the *Monatshefte für Mathematik und Physik* received a twenty-five-page paper written by Kurt Gödel, a logician working in Vienna, who was the first to demonstrate that certain mathematical statements can neither be proved nor disproved. In effect, Gödel showed the inevitability of finding in arithmetic logical paradoxes that are the equivalent of the statement 'This statement is unprovable.' In other words, if it is provable, it is false, and mathematics is therefore inconsistent. And if it's unprovable, then it is true and mathematics is incomplete.

To make matters worse, Gödel also showed that it is never possible to prove that a mathematical system itself is logically self-consistent. One must always step outside a formal mathematical calculus to determine its validity. This is an explosive finding. As if to dramatise the limited power of pure logic, Gödel himself became so paranoid in later life that he starved himself to death.

For physicists engaged in the quest for a Theory of

Everything, Gödel's work has the shocking implication that we might never be able to prove we have found one. This seems to relegate the mathematical foundation of science to the status of just another religion, says John Barrow of Cambridge University: 'If we were to define a religion to be a system of thought which contains unprovable statements, so it contains an element of faith, then Gödel has taught us that not only is mathematics a religion but it is the only religion able to prove itself to be one.' What did Gödel think were the implications of his work? Somewhat surprisingly, he considered that mathematical intuition, by which we can 'see' the truths of mathematics and science, would one day be valued as highly as logic itself.

His research was not the only blow to Hilbert's dream. In 1936, the great British mathematician and father of the computer, Alan Turing, took the effort started by Gödel's 'incompleteness theorem' another step forward by discovering the concept of uncomputability. First Turing came up with the revolutionary notion of a general-purpose or universal computer. And then he immediately showed that there are limits to what such a machine can do.

This unsettled mathematicians who usually think that they 'read God's thoughts', comments Greg Chaitin of the IBM Thomas J. Watson Research Laboratory. Mathematicians believe that they deal with certainties, while ordinary mortals deal with doubts. 'Certainly, a mathematical proof gives more certainty than an argument in physics or than experimental evidence, but mathematics is not certain. This is the real message of Gödel's famous incompleteness theorem and of Turing's work on uncomputability.'

Chaitin continued this effort to probe the limits of mathematics and began to suspect that incompleteness could reflect the fact that there is randomness lurking at the heart of pure mathematics. During the 1980s, he discovered that even the simplest version of arithmetic – which uses only whole numbers such as 1, 2 and 100 – contains intrinsic randomness. He found a number, called Omega, with a rather magical property: although it is perfectly well defined mathematically it can, alarmingly, never be written down. Every one of its digits has to be from 0 to 9 but that is about all we do know. Each digit is random. 'Gödel's theorem, Turing's work and my own results show that even in math, you can't know the whole truth and nothing but the truth.'

The Omega number puts magic at the heart of mathematics. The number transcends human ability and, if he were around, Einstein probably would have said that only God could know the value of Omega. 'You have no pattern, indeed complete chaos,' says Chaitin. 'Surprisingly, even arithmetic possesses random elements. God not only plays dice in quantum mechanics, but even with the whole numbers.'

Many have pondered the implications of Gödel's, Turing's and Chaitin's discoveries. Some, like the scientist Freeman Dyson, take an optimistic view: if it is impossible to sum up all mathematics in a set of axioms (rules), then science is an endless quest to find new sets of axioms to describe features of the universe – perhaps it would take an infinite number of axioms to capture its full richness. Chaitin himself believes that the arithmetical randomness that he discovered is a feature of the Platonic world of mathematical possibilities, not the real one that we inhabit. Another optimist is Barrow, who points out that mathematical structures that are simpler than arithmetic (such as geometry, or mathematics in which only the plus and minus operations are used) are immune to Gödel's and Turing's theories. Perhaps, Nature, like a computer, runs with a decidable logic which is smaller than the whole of arithmetic. 'The mathematics that Nature makes use of may be smaller and simper than is needed for these problems to arise,' Barrow says.

But can we be sure? Pessimists, like the writer on theology and science, Stanley Jaki, suggest that because science is based on mathematics, and mathematics cannot discover all truths, therefore science cannot discover all truths. Others have found evidence of uncomputability in quantum gravity, in differential equations that are in common use in physics, and in attempts to produce a mathematical account of an equilibrium between competing influences. The mathematical physicist, Stephen Wolfram, has speculated that the limitations found by Turing could be common and has also found examples in physics. Although he believes the wonders of our universe can be captured by simple rules, Wolfram states in his vast tome *A New Kind of Science* that very often there can be no other way to know the consequences of these rules other than to watch and see how they unfold. He calls this 'computational irreducibility' and he believes that all phenomena, whether the actions of wizards or the phenomena found in nature, can be viewed as computations and rather simple ones at that. Most of them cannot be cast in the form of equations that can get to the end result of the computation faster than the computation itself. The traditional mathematics used in science is too limited to capture the full complexity of the cosmos, whether Harry Potter's thought processes or the turbulence in a boiling cauldron.

In this strange sense, there is something occult lurking in the mathematical foundations of science, a profound puzzle that should be investigated by Bode and Croaker, the Unspeakables from the wizard world's Department of Mysteries. Einstein once remarked that the most incomprehensible thing about the universe is that it is comprehensible. Given the above limits of mathematics, can we be sure?

Physicists still dream of what they call a final theory, or a theory of everything. Some, like the cosmologist Stephen Hawking, believe we may even have identified that theory: what is known in the scientific world as M theory, where M stands for 'mystery', 'membrane' or the 'mother of all strings', which has been under construction by a number of physicists worldwide.

Perhaps the M stands for magic. One of the strangest features of such theories is that they require the universe to have more than three spatial dimensions to unify our picture of all forces and all matter. M theory goes beyond the four dimensions we live in – three of space and one of time – to suggest that there are as many as eleven. This 'hyperspace' is impossible to visualise in one go and, once again, it is no wonder the public finds such work intimidating. All that matters is whether it can accurately model the behaviour of the real world. And in the case of M theory, we may be close to providing its first real-world test.

While experiments have highlighted cracks in the current best theory, called 'the standard model', Hawking believes that M theory's extra hidden dimensions could be revealed in 2006, when the biggest machine on the planet – a Geneva atom smasher called the Large Hadron Collider – is ready for scientists to run experiments.

However, even if the work of Chaitin, Turing and Gödel proves irrelevant (as everyone hopes), and even if M theory is confirmed by endless experiments, achieving a theory of everything will not remove the magic and mystery from the universe – that is, if James Hartle of the University of California, Santa Barbara, turns out to be correct.

To use a theory of everything, scientists will have to incorporate a theory about the conditions that got everything going in the first place, the first 'point' of the trajectory that brought the universe to its current state. Even if the work of Stephen Hawking and others on this problem pays off so that we know these initial conditions, Hartle points out that the theory of everything could never predict that a black cat evolved, or that it shrieks meow when dropped from a window (even though that theory would tell us with great accuracy how fast the feline would accelerate).

There are profound limitations to what this theory could do. There are not enough protons in the universe to make the compact disks that would be necessary to describe the cosmos at any instant, down to the level of one millimetre, Hartle estimates. This vast stack of information could not be summed up by a theory of everything, which compresses such information by looking for regularities, such as those described by the laws of physics. We can only discover laws that are simple enough to discover. And, 'If it is short enough to be discoverable it is going to be too short to predict everything,' insists Hartle.

Among those laws is quantum mechanics, which describes everything in terms of probabilities (so that the point on the trajectory referred to the paragraph before last would actually be couched in terms of probabilities). The good news is that quantum theory really can predict the movement of the stock market. The bad is that such predictions will mostly be meaningless fifty-fifty odds. Hartle points out that the results of evolution, such as cats (and presumably witches), have much more to do with quantum accidents over the course of history than the fundamental laws of nature.

Think of the fluctuations that led to the formation of our galaxy and then that of the solar system. Then came that happy accident that led to the first primitive unicellular life on earth. From there, cats, humans and wizards appear after a long chain of accidental mutations over the course of four billion years. 'No one is going to come along with a nasty bunch of equations and explain why you like vanilla ice-cream. It isn't going to happen,' says Hartle. 'The regularities of human history, psychology, economics and biology and so forth obey the fundamental laws. But they do not follow from them.'

The universe will remain safe for psychologists, biologists, historians and other environmental scientists. A theory of everything is not a theory of everything. It does not explain all features of the world. It would have a minimal impact on most science. A physical theory of everything might be irrelevant at the level of biological information or human thought, at the level of reality which matters to us, just as the to-and-fro of electrons in a PC's microchips are invisible to and unheeded by a computer operator. Everything is indeed governed by Schrödinger's equation, a quantum expression that can describe all atoms or groups of atoms. However, you can't readily express a babbling brook, a thought on a wing or a crackling electronic logic circuit in terms of atoms, and it is at this length scale, between the cosmic and atomic domains, where the most challenging science lies.

A theory of everything would also be silent on issues that concern many of us. Does God love us? Does life have any meaning? It would not provide a God's-eye view of creation that will vanquish our sense of mystery or slake our thirst for the truth that is out there or our yearning for that something that is beyond our experience. Magic will persist.

Why we have faith in science

In the light of these profound limitations, science does not seem so special. Moreover, like magic, science is just another human endeavour. While the final results of scientific investigation may be cold, logical and impersonal, the process is not. Being human, scientists themselves can be competitive, boastful, sly and deceitful. Science is performed by people who, like any witch or wizard, can be prejudiced, make mistakes and jump on passing bandwagons. False theories, complacent conservatism and fraudulent claims can thrive within science, as they do in any other human pursuit.

Science does, however, at least attempt to account for and do away with subjectivity through experiment, and not just one but many. Confirmation is crucial. Although many complain that science is too closed to novel ideas, there are many examples of how, when a stack of evidence mounts against established thinking, those views are abandoned or modified. These examples range from quantum theory, which made Einstein uncomfortable, to the idea of infectious proteins (prions) which were once derided as a biological heresy.

One of the characteristic features of magical thought and religious faith that makes them so different from science is that, once the initial premises are accepted, no subsequent discovery will necessarily break the believer's faith or belief, for he can often find a way to explain it away. When a magical spell does not work, it may say more about the person muttering it than the spell itself. Perhaps, like Ron Weasley's attempts at making a feather fly, the magical words were not pronounced properly. (It is 'Wing-*gar*-dium Levi-*o*-sa', as Hermione corrected him.)

Some sociologists and philosophers – under the guise of relativism – argue that knowledge is socially conditioned and culturally determined; there is no one single truth about the external world. All beliefs are equally valid, and scientific truth, being one of them, is an illusion.

Science does indeed have its own beliefs, such as that the laws of physics are universal and that symmetry plays a profound role in shaping elegant mathematical theories of the universe by helping to simplify calculations. And, as mentioned, above, scientists also believe that the behaviour of the universe can be mapped onto mathematics.

Scientists also come up with elaborations to explain away the shortcomings of a dud theory, just as sorcerers conjure up excuses for the shortcomings of a dud spell. But unlike other belief systems, those of science are universal and culture-free because they are endlessly sifted by experiment Science will eventually abandon any belief or 'truth' if the evidence requires it.

Science really is special. Science really is the best way of understanding how the world works. Unlike technology or religion, it only originated once in history, in ancient Greece. Even if history were rerun and it took a different course, the conclusions of science would be the same – DNA would still be the genetic material of inheritance, hydrogen would still be the most common element in the universe, and stars would still be powered by nuclear fusion.

If Newton had not, as Wordsworth put it, voyaged through strange seas of thought alone, someone else would have. If Marie Curie had not lived, we still would have discovered the radioactive elements polornium and radium. But if J.K. Rowling had not been born, we would never have known about Harry Potter. That is why Master Potter means so much to all of us. Science may be special but Harry, as a work of art, is more so. Harry Potter is unique.

References

Aldhouse-Green, Stephen, *Paviland Cave and the Red Lady* (Western Academic & Specialist Press, Bristol, 2000)

Armbruster, Peter, 'Heavy ion fusion and exotic nuclei studies at SHIP', *The Neutron and Its Applications*, ed. P. Schofield, Conference Series Number 64 (Bristol and London, The Institute of Physics, 1982)

Bahn, Paul, and Jean Vertut, *Journey Through the Ice Age* (London, Seven Dials, 1999)

Balaban, E., M.-A. Teillet and N. LeDouarin, 'Application of the quail-chick chimeric system to the study of brain development and behavior', *Science* 241:1339–42 (1990)

Balaban, E., 'Changes in multiple brain regions underlie species differences in a complex, congenital behavior', *Proceedings of the National Academy of Sciences* (USA), 94:2001–6 (1997)

Balick, Michael, and Paul Alan Cox, *Plants, People and Culture: The Science of Ethnobotany* (New York, Scientific American Library, 1997)

Barrow, J., 'Mathematical jujitsu: some informal thoughts about Gödel and physics', *Complexity*, Vol. 5 (issue 5), 28–34 (2000)

Berry, M.V., 'The Levitron and adiabatic trap for spins', *Proceedings of the Royal Society London*, A 452, 1207–20 (1996)

Berry, M.V., and A.K. Geim, 'Of flying frogs and Levitrons',

European Journal of Physics, 18, 307–13 (1997)
Blackmore, Susan, *The Meme Machine* (Oxford, Oxford University Press, 1999)
Bouchard, Thomas, Matt McGue, David Lykken and Auke Tellegen, 'Intrinsic and extrinsic religiousness: genetic and environmental influences and personality correlates', *Twin Research*, Vol. 2, 88–98 (1999)
Braun, Kathryn, Rhiannon Ellis and Elizabeth Loftus, 'Make my memory: how advertising can change our memories of the past', *Psychology & Marketing*, Vol. 19 (1), 1–23 (2002)
Brookes, Martin, *Fly: An Experimental Life* (London, Weidenfeld & Nicolson, 2001)
Caporael, Linnda, 'Ergotism: the Satan loosed in Salem?', *Science*, 192, 21–6 (1976)
Chaitin, Gregory, *Conversations with a Mathematician* (London, Springer, 2002)
Cheyne, Allan, 'The ominous numinous sensed presence and "other" hallucinations', *Journal of Consciousness Studies*, 8, 5–7 (2001)
Cheyne, Allan, Ian Newby-Clark and Steve Rueffer, 'Relations among hypnagogic and hypnopompic experiences associated with sleep paralysis', *Journal of Sleep Research*, 8, 313–17 (1999)
Cheyne, Allan, Ian Newby-Clark and Steve Rueffer, 'Hypnagogic and hypnopompic hallucinations during sleep paralysis: neurological and cultural construction of the nightmare', *Consciousness and Cognition*, 8, 319–37 (1999)
Colbert, David, *The Magical Worlds of Harry Potter: A Treasury of Myths, Legends and Fascinating Facts* (London, Puffin, 2001)
Collins, Martin, 'A female giant squid (Architeuthis) stranded on the Aberdeenshire coast', *Journal of Molluscan Studies*, 64, 489–92 (1998)
Davis, W., *Passage of Darkness: The Ethnobiology of the Haitian Zombie* (Chapel Hill, University of North Carolina Press, 1988)

Favreau P., N. Gilles, H. Lamthanh, R. Bournaud, T. Shimahara, F. Bouet, P. Laboute, Y. Letourneux, A. Menez, J. Molgo and F.A. Le Gall, 'New omega-conotoxin that targets N-type voltage-sensitive calcium channels with unusual specificity', *Biochemistry*, Vol. 40, No. 48, 14567–75 (2001)

Fehr, E., and S. Gächter, 'Altruistic punishment in humans', *Nature*, Vol. 415, No. 6868, 137–40 (2002)

Fink, Thomas, and Yong Mao, *The 85 Ways to Tie a Tie* (London, Fourth Estate, 1999)

ffytche, D.H., and R.J. Howard, 'The perceptual consequences of visual loss: positive pathologies of vision', *Brain*, Vol. 122, 1247–60 (1999)

Fuente-Fernández, Raúl de la, Thomas J. Ruth, Vesna Sossi, Michael Schulzer, Donald B. Calne and A. Jon Stoessl, 'Expectation and dopamine release: mechanism of the placebo effect in Parkinson's disease', *Science*, Vol. 293, No. 5532, 1164–6 (2001)

Fukuda, Kazuhiko, Robert Ogilvie, Lisa Chilcott, Ann-Marie Vendittelli and Tomoka Takeuchi, 'The prevalence of sleep paralysis among Canadian and Japanese college students', *Dreaming*, Vol. 8, No. 2, 59–66 (1998)

Fukuda, Kazuhiko, Robert Ogilvie and Tomoka Takeuchi, 'Recognition of sleep paralysis among normal adults in Canada and Japan', *Psychiatry and Clinical Neurosciences*, Vol. 54, 292–3 (2000)

Furst, P., *The Encyclopaedia of Psychoactive Drugs. Mushrooms: Psychedelic Fungi* (Burke Publishing, London, 1986)

Gerloff, Sabine, '*Bronzezeitliche Goldblechkronen aus Westeuropa. Betrachtungen zur Funktion der Goldblechkegel vom Typ Schifferstadt und der atlantischen "Goldschalen" der Form Devils Bit and Axtroki*', in A. Jockenhövel (ed.), Festschrift für Hermann Müller-Karpe zum 70, Geburtstag (Dr Rudolf Habeltin, Bonn, 1995)

Gomez-Alonso, J., 'Rabies: a possible explanation for the vampire legend', *Neurology*, 51(September 1998), 856–9

Gregory, Richard, John Harris, Priscilla Heard and David Rose, *The Artful Eye* (Oxford, Oxford University Press, 1995)
Gurdon, John, and P.Y. Bourillot, 'Morphogen gradient interpretation', *Nature*, Vol. 413, 797–803 (2001)
Guthrie, Edwin, and George Horton, *Cats in a Puzzle Box* (New York: Rinehart, 1946)
Haggard, Patrick, Sam Clark and Jeri Kalogeras, 'Voluntary action and conscious awareness', *Nature Neuroscience*, 5, 382–5 (2002)
Harding, Rosalind, 'Interpreting patterns of diversity in the melanocortin 1 receptor gene', VIIIth CEPH Annual Conference, Paris, France, 25–26 May 2000
Holmes, Michael, 'Revolutionary birthdays', *Nature*, Vol. 373, No. 6514, 468 (1995)
Huffman, Michael, 'Self-medicative behavior in the African great apes: an evolutionary perspective into the origins of human traditional medicine', *BioScience*, Vol. 51, No. 8, 651–61 (2001)
Hutton, R., *The Pagan Religions of the Ancient British Isles* (Oxford, Blackwell, 2000)
Hyde, Peter, and Eric Knudsen, 'The optic tectum controls visually guided adaptive plasticity in the owl's auditory space map', *Nature*, Vol. 415, No. 6867, 73–6 (2002)
Julsgaard, B., A. Kozhekin and E.S. Polzik, 'Experimental long-lived entanglement of two macroscopic quantum objects', *Nature*, Vol. 413, No. 6854, 400–3 (2001)
Koenig, Harold, *Is Religion Good for Your Health? Effects of Religion on Physical and Mental Health* (Haworth Press, New York, 1997)
Koenig, H.G., H.J. Cohen, D.G. Blazer et al, 'Religious coping and depression in elderly hospitalized medically ill men', *American Journal of Psychiatry*, Vol. 149, 1693–700 (1992)
Koenig, H.G., S. Ford, L.K. George, D.G. Blazer and K.G. Meador, 'Religion and anxiety disorder: an examination and comparison of associations in young, middle-aged, and eld-

erly adults', *Journal of Anxiety Disorders*, Vol. 7, 321–42 (1993)

Knudsen, E.I., and P.F. Knudsen, 'Disruption of auditory spatial working memory by inactivation of the forebrain archistriatum in barn owls', *Nature*, Vol. 383, No. 6599, 428–31 (1996)

Kristensen, Reinhardt, and Peter Funch, 'Phylum Cycliophora', in the *Atlas of Marine Invertebrate Larvae*, ed. Craig Young, Mary Sewell and Mary Rice, 199–208 (Academic Press, 2002)

Kronzek, Allan, and Elizabeth Kronzek, *The Sorcerer's Companion: A Guide to the Magical World of Harry Potter* (New York, Broadway, 2001)

Kurtén, Björn, *Dance of the Tiger: A Novel of the Ice Age* (Berkeley, University of California Press, 1995)

Lamont, Peter, and Richard Wiseman, *Magic in Theory* (Hatfield, University of Hertfordshire Press, 1999)

Lenhoff, Howard, 'A real-world source for the "little people": the relationship of fairies to individuals with Williams Syndrome', *Proceedings of the Eaton Symposium*, ed. G. Slusser (Athens, Georgia, University of Georgia Press, 1999)

Littlewood, Richard, and Chavannes Douyon, 'Clinical findings in three cases of zombification', *Lancet*, Vol. 350, 1094–6 (1997)

Loftus, Elizabeth, and William Calvin, 'Memory's future', *Psychology Today*, 34 (2): 55ff (March–April 2001)

May, R., 'The ecology of dragons', *Nature*, Vol. 264, No. 5581, 16–17 (1976)

Mayor, Adrienne, *The First Fossil Hunters* (Princeton, Princeton University Press, 2001)

Mayor, Adrienne and W. Sarjeant, 'The folklore of footprints in stone: from classical antiquity to the present', *Ichnos*, Vol. 8, No. 2, 143–63 (2001)

Mann, John, *Murder, Magic and Medicine* (Oxford, Oxford University Press, 2000)

Mazzoni, G., Elizabeth Loftus and Irving Kirsch, 'Changing beliefs about implausible autobiographical events. A little plausibility goes a long way', *Journal of Experimental Psychology: Applied*, Vol. 7, No. 1, 51–9 (2001)

McGowan, Christopher, *The Dragon Seekers: The Discovery of Dinosaurs During the Prelude to Darwin* (London, Little, Brown, 2002)

Mizukami, Yukiko, and Robert Fischer, 'Plant organ size control: Aintegumenta regulates growth and cell numbers during organogenesis', *Proceedings of the National Academies of Science*, Vol. 97, No. 2, 942–7 (2000)

Millis, Mark, 'Harry Potter and NASA?' at www.spacekids.com (1 August 2000)

Moore, Bruce, and Susan Stuttard, 'Dr Guthrie and *Felis domesticus*. Or: Tripping over the cat', *Science*, Vol. 205, No. 4410, 1031–3 (1979)

Mueller, Paul, David Plevak and Teresa Rummans, 'Religious involvement, spirituality, and medicine: implications for clinical practice', *Mayo Clinic Proceedings*, 76, 1225–35 (2001)

Nelson, G., M.A. Hoon, J. Chandrashekar, Y. Zhang, N.J. Ryba and C.S. Zuker, 'Mammalian sweet taste receptors', *Cell*, August 10, 106 (3), 381–90 (2001)

Nelson G., J. Chandrashekar, M. Hoon, L. Feng, G. Zhao, N. Ryba and C. Zuker, 'An amino-acid taste receptor', *Nature*, Vol. 416, No. 6877 199–202 (2002)

Pargament, Kenneth, *The Psychology of Religious Coping: Theory, Research, and Practice* (New York, Guilford Press, 1997)

Pavlidis, Ioannis, Norman Eberhardt and James Levine, 'Seeing through the face of deception', *Nature*, Vol. 415, No. 6867, 35 (2002)

Peña, J.L., and M. Konishi, 'Auditory spatial receptive fields created by multiplication', *Science*, 292, 249 (2001)

Petrovic, Predrag, Eija Kalso, Karl Magnus Petersson and Martin Ingvar, 'Placebo and opioid Aralgesia – imaging a

shared neuronal network', *Science*, Vol. 295, No. 5560, 1737–40 (2002)
Povinelli, Daniel, *Folk Physics for Apes: The Chimpanzee's Theory of How the World Works* (Oxford, Oxford University Press, 2000)
Rakkolainen, I., and K. Palovuori, 'WAVE – A walk-thru virtual environment', IEEE VR 2002 Conference, Proceedings of Immersive Projection Technology Symposium, Orlando, Florida, USA, 24–25 March 2002
Ramachandran, V. S., 'Blind spots', *Scientific American*, 266 (5), 44–9 (1992)
Ramachandran, V. S., and R.L. Gregory, 'Perceptual filling in of artificially induced scotomas in human vision', *Nature*, 350, No. 6320, 699–702 (1991)
Richman, J., S.-H. Lee, K. Fu and J. Hui, 'Noggin and retinoic acid transform the identity of facial avian prominences', *Nature*, Vol. 414, No. 6866, 909–12 (2001)
Riddle, John, and Worth Estes, 'Oral contraceptives in ancient and medieval times', *American Scientist*, Vol. 80, 226–33 (1992)
Rowling, J.K., *Harry Potter and the Philosopher's Stone* (London, Bloomsbury, 1997)
Rowling, J.K., *Harry Potter and the Chamber of Secrets* (London, Bloomsbury, 1998)
Rowling, J.K., *Harry Potter and the Prisoner of Azkaban* (London, Bloomsbury, 1999)
Rowling, J.K., *Harry Potter and the Goblet of Fire* (London, Bloomsbury, 2000)
Sagan, Carl, *Contact* (London, Arrow Books, 1987)
Santhouse, A.M., R.J. Howard and D.H. ffytche, 'Visual hallucinatory syndromes and the anatomy of the visual brain', *Brain*, Vol. 123, 2055–64 (2000)
Scamander, Newt, *Fantastic Beasts and Where to Find Them* (London, Obscurus Books, 2001)
Schafer, Elizabeth, *Exploring Harry Potter* (London, Ebury, 2000)
Shams, L., Y. Kamitani and S. Shimojo, 'What you see is what

you hear', *Nature*, Vol. 408, No. 6814, 788 (2000)

Shapiro, Beth, Dean Sibthorpe, Andrew Rambaut, Jeremy Austin, Graham M. Wragg, Olaf R.P. Bininda-Emonds, Patricia L.M. Lee and Alan Cooper, 'Flight of the dodo', *Science*, Vol. 295, No. 5560, 1683 (2002)

Simons, Daniel, and Daniel Levin, 'Failure to detect changes to people during real-world interaction', *Psychonomic Bulletin and Review*, Vol. 4, 644–9 (1998)

Skinner, B.F., '"Superstition" in the pigeon', *Journal of Experimental Psychology*, 38, 168–72 (1948)

Stone, Trevor, and Gail Darlington, *Pills, Potions and Poisons: How Drugs Work* (Oxford, Oxford University Press, 2000)

Stouffer, Nancy, *The Legend of Rah and the Muggles* (Unpublished, 1984)

Surette, M.G, M.B. Miller and B.L. Bassler, 'Quorum sensing in *Escherichia coli*, *Salmonella typhimurium* and *Vibrio harveyi*. A new family of genes responsible for autoinducer production', *The Proceedings of the National Academy of Sciences*, 96, 4, 1639–44 (1999)

Tacon, Paul, and Christopher Chippindale, 'Transformations and depictions of the First People: animal-headed beings of Arnhem Land, N.T. Australia', in *Theoretical Perspectives in Rock Art Research*, ed. Knut Helskog (Oslo, Instituttet for sammenlignende kulturforskning, 2001)

Taylor, Timothy, 'The Gundestrup cauldron', *Scientific American*, Vol. 266, No. 3, 84–9 (1992)

Taylor, Timothy, and Paul Budd, 'The faerie smith meets the bronze industry: magic versus science in the interpretation of prehistoric metal making', *World Archaeology*, Vol. 27, No. 1, 133–43 (1995)

Taylor, Timothy, *The Prehistory of Sex* (London, Fourth Estate, 1996)

Taylor, Timothy, 'Explanatory tyranny', *Nature*, Vol. 411, No. 6836, 419 (2001)

Visalberghi, E., and M. Tomasello, 'Primate causal understanding

in the physical and in the social domains', *Behavioral Processes*, 42, 189–203 (1998)
Visick, Karen, and Margaret McFall-Ngai, 'An exclusive contract: specificity in the *Vibrio fischeri–Euprymna scolopes* partnership', *Journal of Bacteriology*, Vol. 182, No. 7, 1779–87 (2000)
Vollrath Fritz, 'To catch a fly', *Science and Public Affairs* (4), 20–24 (1994)
Vyse, Stuart, *Believing in Magic: The Psychology of Superstition* (Oxford, Oxford University Press, 1997)
Vyse, Stuart, 'Believing in magic: the psychology of superstition', *The General Psychologist*, Vol. 34, No. 3, 71–7 (1999)
Westwood, Jennifer, *Albion: A Guide to Legendary Britain* (London, Guild Publishing, 1986)
Whisp, Kennilworthy, *Quidditch Through the Ages* (London, Whizz Hard Books, 2001)
Whittington, Michael, *The Sport of Life and Death: The Mesoamerican Ballgame* (London, Thames & Hudson, 2001)
Wolfram, Stephen, *A New Kind of Science* (Champaign, Illinois, Wolfram Media, 2002)
Wolpert, L., *The Triumph of the Embryo* (Oxford, Oxford University Press, 1991)
Wolpert, L., *The Unnatural Nature of Science* (London, Faber and Faber, 1992)
Wooley, Benjamin, *The Queen's Conjurer: The Science and Magic of Dr Dee* (London, HarperCollins, 2001)
Zald, David, Dorothy Mattson and José Pardo, 'Brain activity in ventromedial prefrontal cortex correlates with individual differences in negative affect', *Proceedings of the National Academy of Sciences*, Vol. 99, No. 4, 2450–4 (2002)
Zeki, Semir, *Inner Vision: An Exploration of Art and the Brain* (Oxford, Oxford University Press, 1999)

Glossary of Muggle science, Potter magic, oddments and tweaks

Absolute zero No winter at Hogwarts can be as cold as this. The lowest temperature that anything can approach is minus 273.15 degrees C, known as absolute zero. (For comparison, a deep-freezer is around minus 15 degrees C.) A fundamental law, known as the Third Law of Thermodynamics, says that nothing can be cooled down to this temperature. As something cools, the average energy and movement of its molecules drops. You might think they would come close to stopping altogether near absolute zero. As ever, this commonsense view is overturned by quantum theory which says that even at such extremely low temperatures, molecules cannot help but jiggle about a bit.

Alkaloids Plant chemicals, such as morphine, strychnine and quinine, which contain nitrogen and often occur in seeds at high concentrations, where they are stored as a source of nitrogen for the growing plant. A high proportion of alkaloids offer protection against insects – they stimulate what is termed 'deterrent' nerve cells in the taste hairs (sensilla) used by insects to select food.

Amino acids The molecular building blocks of proteins.

Amulets Protective devices worn around the body or placed next to other objects to protect against various evils. They range from a red string tied around the wrist and a stone carried in a small pouch around the neck to a piece of iron tied to one's bed. One papyrus amulet, written in Greek, was folded, rolled, and carried in a tube to protect the bearer from the onslaught of fever.

Aragog A giant spider found in the Forbidden Forest. Fritz Vollrath of Oxford University, one of the world's leading spider experts, has worked out that the webs spun by even a human-sized garden spider would be formidable: they would be the size of a football field, able to catch a jumbo jet at landing speed and would fit into a tea chest when folded up. Far from being weak, spider silk is in fact extremely strong, ounce for ounce much stronger than nylon or steel. Silk is not only strong but also very elastic. Some spider silks, especially the sticky silk of the garden spider's orb web, are much more elastic than any commercial rubber.

Atoms Atoms were thought by the ancient Greeks to be indivisible units of matter. Now they are seen as the smallest units that bear the chemical characteristics of an element, whether hydrogen or uranium. Around two hundred thousand million million would fit on the full stop at the end of this sentence. They are mostly empty space: the nucleus, where most mass resides, is 100,000 times smaller than the overall atom. Each atom consists of a positively charged nucleus orbited by a mist of negative charge.

Bearded ghoul A spiky and strange-looking poisonous fish found in the waters off Australia.

Beauxbatons horses These winged horses, each the size of an elephant, drink single-malt whisky. No doubt they are generously endowed with enzymes to handle alcohol and its breakdown product, acetaldehyde.

Beer An alcoholic drink produced by yeast fermentation of a solution of sugar derived from cereals, such as malted barley. The recipe for butterbeer, the favourite drink of Barny the Fruitbat, remains a mystery.

Belladonna Renaissance women dropped the juice of belladonna (Italian for 'beautiful woman') berries into their eyes to dilate the pupils and make them more attractive.

Bicorn Refers to either having two horns or, more interestingly, a mythical beast that grew fat through living on good and enduring husbands.

Black dog Harry was haunted in his third adventure by one – a portent of death called the Grim (which turned out to be an Animagus). Many spectral hounds supposedly prowl around Britain, according to Muggle folklore. The Trash or Skriker is found in Lancashire, while the Mauthe Doog is found on the Isle of Man. In Suffolk, the hound is known as Shuck, Old Shuck, or the Shuck dog. There were also packs of spectral hounds, notably the Dandy Dogs in Cornwall and Gabriel Ratchets in the north. They often haunted bridges, crossroads, gateways and footpaths that were thought to be transition points between the mortal world and that of the supernatural. There is no reference to one in Magnolia Crescent.

Boarhound A large dog, such as the Great Dane, used

originally for hunting wild boars.

Boggart Shape-shifter that likes enclosed spaces and can adopt the appearance of whatever it thinks will frighten you the most. Repelled by laughter. Muggles in the north of the UK use this term for ghosts and poltergeists. One that haunted Boggart Hole Clough, in the northwest, clattered about in clogs.

Boomslang Large greenish venomous snake found in South Africa (from the Afrikaans: *boom*, tree, and *slang*, snake).

Boring tie Mr Dursley likes to pick the most boring tie for work. Fortunately for him, mathematics has shown that it takes more than a dull pattern to make a dull tie. A mathematical analysis by Thomas Fink and Yong Mao has identified that there are eighty-five different ways to tie a tie. The most boring is probably the simplest: the Oriental, which begins with the tie inside-out around the neck. It is also known as the simple knot, red knot and *petit noeud*.

Cell A discrete, membrane-bound portion of living matter, the smallest unit capable of an independent existence.

Centaur Human–animal composites such as centaurs (man/horse), satyrs (man/goat) and tritons (man/fish) were reported by the ancients but they are likely to be hoaxes, along the lines of modern sightings of Bigfoot, ET and mermaids, according to the independent researcher Adrienne Mayor. In the 1980s, an ancient centaur skeleton from northern Greece toured America – but it was a fake created by William Willers, a zoology professor and artist, to 'gain some insight into the psychology of the ancients'.

Chaos A term often used to describe what ensues if electric-blue Cornish pixies run amok. In science, it is used to describe apparently random behaviour. The essence of chaos is expressed in the butterfly effect: a butterfly flapping its wings near Hogwarts can cause a subsequent hurricane over another school of wizardry located far away, such as Beauxbatons. This concept put paid to the prospect of truly long-range weather forecasting. Earth's atmosphere is so 'sensitive' that if there is even the slightest uncertainty in the current weather conditions then the weather in a few weeks' time, such as for the Quidditch World Cup, is unpredictable. Dumbledore understands our blinkered picture of the future only too well and remarks on how the consequences of our actions are always so complicated and diverse that predicting the future is fraught with difficulties.

Chocolate The treatment for any encounter with soul-sucking Dementors – chocolate – is a fatty psychoactive food made from the seeds of the *Theobroma cacao* tree. The use of chocolate dates back thousands of years to when it was first cultivated in Central America, including what today is Guatemala, Mexico and Belize. Archaeologists tell us that chocolate was a key part of elite Mayan culture, which had a special cacao god, and the Aztecs believed that drinking chocolate would bring wisdom, understanding and energy. Chocolate contains caffeine, and methylxanthine and theobromine, both caffeine-like substances. Although they do not act on the brain as strongly as caffeine, they do stimulate heart muscle. More provocative is the discovery that chocolate contains a range of 'drugs' that can act on the brain. No single one of them is present in particularly high concentrations in chocolate, but the combination may have an effect. One is phenylethylamine, a messenger chemical (neurotransmitter) naturally found in the brain that raises levels of another neurotransmitter, dopamine, in pleasure centres, boosting blood pressure and heart rate while heightening sensation. Neuropharmacologists have even found marijuana-like substances in chocolate, substances from the N-acylethanolamine group of chemicals. Studies on rats suggest that the wonderful taste of chocolate can boost the release of natural opiates – morphine-like chemicals – called endorphins. But while some claim that chocolate's chemistry, precise melting point and neuropharmacology explain its attractions, others cite powerful psychological factors. Nature has conditioned us to crave sweet, fatty food, and chocolate represents an ideal combination of the two.

Chromosome A long strand of DNA, containing thousands of genes, in a package. There are twenty-three pairs in all human cells, except eggs and sperm.

Constellations The Latin names for the constellations include a number that will sound familiar to readers of Harry Potter: Centaurus the Centaur; Draco the Dragon; Lupus the Wolf; Monoceros the Unicorn; Phoenix the, er, Phoenix; and Vela, the Sail of the Argonauts' ship.

Cruciatus curse The crucifying pain caused by this curse originates in the body but is perceived by the brain, according to Qasim Aziz of Hope Hospital in Salford, where a new brain-scanning method has revealed the origins of an 'ouch'. Pain is first perceived in the sensory cortex, which is sometimes referred to as the homunculus: all parts of the body are represented on this part of the brain, which lies directly under a band that stretches from ear to ear. The foot and lower limbs are at the top, while the mouth and lips and gut areas are on the lower parts, just above the ears. As the intensity of pain increases, it would place greater cognitive and emotional demands. Brain scans would reveal this as increased activity in the anterior insula, at the back of the ears, and within the deeper anterior cingulate cortex. In addition, crucifying pain would stimulate activity in the prefrontal cortex, just behind the forehead, which is involved in making decisions and judgements about a predicament.

Currency Each of Harry's galleons stored in Gringotts Bank is worth 17 sickles and each sickle is worth 29 knuts. That means, of course, that a galleon is worth 493 knuts.

Curses Lead tablets known in Greek as *katadesmoi* and in Latin as *defixiones* were used for cursing and binding rivals, whether a suitor who pursued the same love interest or a competitor in sports or business. They appear to have originated in Greece in the fifth century BC and spread from there throughout the Mediterranean world. Initially, a victim's name was scratched on a thin sheet of lead and thrown into graves, pits or wells. Over time, the tablets became more elaborate and their preparation often entailed complex rituals, including the binding, piercing or burning of wax, clay or lead dolls representing the intended victim.

Cyclops According to Greek legend, the Cyclops were a race of terrible giants. Their name, which means 'round eye', refers to a huge eye in the middle of their foreheads. Scientists now suggest that fossil bones of ancient dwarf elephants may have given rise to the myth. The nasal opening in the fossil skulls was probably mistaken for a giant central eye socket by people who did not realise that elephants had once grazed on Mediterranean islands.

Dementors The Dementors, the blind soul-sucking jailers of Azkaban, are the embodiment of the darkest feelings J.K. Rowling experienced during a bout of clinical depression. These creatures drain peace, hope and happiness from their surroundings. As Remus Lupin remarked, even Muggles feel their presence, though they can't see them. Stray too close, and every good feeling and happy memory will be extracted. According to psychiatrist Raj Persaud, the cause of clinical depression remains a puzzle but women are known to be at two to three times greater risk of becoming depressed at some point in life, and the disorder is linked to creativity. Scientists know there is a genetic element from twin and adoption and family studies, and clearly biochemistry plays a role as certain drugs can reliably cause depression as well as cure it. Depression is also linked to life events. Given that depression is so common, perhaps it evolved to allow sufferers to garner sympathy from peers who would then be moved to support rather than continue to compete with them. Effective antidepressants were only discovered serendipitously in the 1950s when it was noted that an antituberculosis treatment made sufferers happier. All antidepressants developed since then work on the principle of changing levels of messenger chemicals such as noradrenaline and serotonin in the brain. But the process remains mysterious; for example, given that the necessary docking sites for the messenger chemicals (receptors) are blocked within a few hours of taking an antidepressant, why does depression tend to respond only a few weeks after the start of treatment? This suggests that actual structural changes to the brain are occurring.

Densaugeo When Malfoy shouted this particular spell at Harry Potter he missed and hit Hermione Granger instead, making her front teeth grow at an alarming rate. Mark Ferguson of the University of Manchester has studied the genes that control tooth growth for many years, and points out that all it takes to make teeth grow longer (and also to erupt faster) is epidermal growth factor, EGF. In his Nobel Prize-winning work, Stanley Cohen administered an extract from the submandibular gland (which was subsequently shown to contain EGF) to newborn mice, which caused their teeth to grow prematurely, or as he put it in his Nobel lecture, 'precocious tooth eruption'.

Determinism The theory that a given set of circumstances inevitably produces the same consequences, rather like the presence of Voldemort and pain in Harry Potter's scar.

DNA (deoxyribonucleic acid) The vehicle of inheritance for all creatures, from Harry to Aragog to Winky. A complex molecule carrying the genetic blueprint for the design and assembly of proteins, the basic building blocks of life. DNA is a ladder-like molecule twisted into a double helix in which the 'rungs' consist of pairs of 'bases' that come in four types (adenine, guanine, cytosine and thymine, or A, G, C and T). The order of the bases provides the blueprint for the proteins. The code is a three-letter one, with a triplet of letters (ATT, say) coding for a particular amino acid that, when joined with a string of others, makes a protein. Harry Potter, like all humans, has about three thousand million bases in his genetic makeup, or genome, a fraction of which are used in the 30,000 genes that work to make proteins.

Dodo The book *Fantastic Beasts* claims that the dodo is not extinct but lives on as the Diricawl, a fluffy bird capable of invisibility. Intriguingly, a DNA analysis of the mummified dodo that helped to inspire *Alice in Wonderland* suggests that in some ways the bird does survive. The dodo turns out to be a glorified pigeon and its closest living relative is the Nicobar pigeon, of Southeast Asia, from which the dodo's common relative diverged about 42.6 million years ago.

Dudley Dursley's bottom Dudley was so obese that his bottom drooped over either side of the kitchen chair and he moaned about the arduous trek he had to make between the television and refrigerator. He piles on the pounds because more energy is taken in than burned by his body – it is as simple as that. Such imbalances are due to both environmental and genetic factors. Environmental factors include how Dudley is spoiled by his parents. However, recent studies have also highlighted genetic factors which include Dudley's predisposition to eat fatty food, such as fried bacon; variations in his resting metabolism (some people seem to burn more calories to stay alive and have a slightly warmer body temperature); activity and exercise levels; and how his body fine-tunes the balance between energy intake and expenditure. Harry, on the other hand, seems to have a genetic complement that enables him to gorge himself at Hogwarts feasts and remain slim.

Enchanted snow Falls warm and dry. This could be

artificial snow consisting of tiny flakes of white polyethylene or biodegradable flakes of modified polylactic acid polymer, from corn or cheese by-products, or from foamed potato and cornstarch.

Entropy A quantity that determines a system's capacity to evolve irreversibly in time, for instance, which describes the propensity of ice lollies to melt and Harry Potter to age. Loosely speaking, we may also think of entropy as measuring the degree of randomness or disorder in a system.

Enzyme The word 'enzyme', which denotes the huge proteins that cells use to transform other molecules, was coined from the Greek words for 'in yeast'. A biological catalyst usually comprised of a large protein molecule consisting of many atoms, even thousands, which accelerates essential chemical reactions in living cells. While a cell has a size of the order of micrometres (millionths of a metre), enzymes and other proteins have a size of several nanometres (billionths of a metre). Ten hydrogen atoms laid side by side can cover a nanometre, which is one-thousandth the length of a typical bacterium and one-millionth the size of a pinhead.

Evolution From the Latin *evolutio*, unfolding. The idea of shared descent of all creatures – humans, goblins and centaurs included. The names of Charles Robert Darwin (1809–82) and Alfred Russel Wallace (1823–1913) have long been joined with the modern concept of 'evolution' and the theory of 'natural selection'. In Darwin's *On the Origin of Species* (1859), he suggested a mechanism: inherited diversity, a struggle for existence that means that not all those born can survive and pass on their heritage; and natural selection (inherited differences in the chance of reproduction). Variants that increase their carrier's ability to make copies of themselves hence become more common; those that hinder it become more rare. In time, Darwin suggested, this led to the evolution of new forms of living organism – the origin of species. Darwin's ideas fit perfectly with those of modern genetics. Diversity arises through mutation, random changes in the genetic material DNA, from generation to generation.

Flash The various colours given out during magic spells can be explained by that revolutionary insight of twentieth-century science, quantum theory (see below). Before quantum theory, it was a puzzle why every atom produces a single definite colour. Why don't some atoms have a little bit of energy spare, and give it up as low-energy (red) light? Why don't others have a lot of energy to release, and give it up as high-energy (blue) light? Quantum theory came up with a solution: atoms have only definite amounts of energy, so can give up only specific amounts and therefore display only specific colours. A yellow flash suggests sodium is around. When any sodium compound, even salt (sodium chloride), is vaporised it forms sodium atoms. If the temperature is high enough, the electrons in the newly formed sodium atoms will not have settled into their lowest energy states. As they do so, they emit the excess energy as radiation, which we see as yellow. Unlike the atomic yellow of sodium, red, green and blue sparkles are formed by molecules and molecular fragments. However, like sodium, these fragments are formed with excess energy and the colours we see are the radiation emitted as they discard this energy. Molecular fragments of strontium and chlorine give red light, of barium and chlorine give yellow-green light, and of copper and chlorine blue light of the kind glimpsed in the dancing flames of the Goblet of Fire.

Flobberworm One candidate for a Muggle Flobberworm came from scientists in the field of evolutionary developmental biology (nicknamed evo devo) comparing the body-plan genes for various creatures to see where the machinery to sculpt heads, arms, legs and so on came from. Sean Carroll of the University of Wisconsin, Madison, Neil Shubin from the University of Pennsylvania and Cliff Tabin of Harvard Medical School deduced that a creature that lived 700 million years ago, dubbed *Urbilateria*, played a crucial role in shaping living things today, from the legs of a supermodel to the wings of a bee. The mythical *Urbilateria* invented a range of body-building genetic machinery so malleable and flexible that it is responsible for the diversity of life on the planet, be it wriggling, crawling, walking or flying.

Fluxweed Used as a remedy for diarrhoea and dysentery.

Fractal geometry From the Latin *fractus*, broken. The geometry used to describe an irregular shape that has the magical property of appearing the same on all scales, whether you look at it close up or far away or in between. Fractals display the characteristic of self-similarity, an unending series of motifs within motifs repeated at all length scales. Fractals abound in nature, for example, clouds, cauliflowers and snowflakes.

Game theory A branch of mathematics and logic that deals with the analysis of games or, more generally, situations where parties have conflicting interests. Principles of game theory also find applications to complicated games such as poker and chess, as well as to real-world problems as diverse as economics, property division, politics and warfare. It can be equally well applied to the attempts at Hogwarts to win the House Cup.

Gene A unit of heredity comprised of the chemical DNA, responsible for passing on specific characteristics from parents to offspring. No research has been conducted to find out if sorcery is linked to genes.

Genetic code The sequence of chemical building blocks of DNA (bases) that spells out the instructions to make amino acids, the building blocks of proteins.

Genetic engineering Tinkering with the genetic code (see DNA) of a creature to produce animals and plants with desirable properties or indeed a host of strange creatures that inhabit Harry's world.

Ghosts Stephen Hawking believes in ghosts, which haunt some of his work on the theory of supergravity, a generalisation of Einstein's theory of gravity that is under development as part of the effort to construct a theory of everything. Sometimes the ghosts are fictitious particles. Sometimes they are particles with negative energy or mass, which is peculiar. Usually they are a sign that the theory is in trouble, though Hawking says, 'One can live quite happily with ghosts.'

Glucose A sugar present in blood that is the source of energy for the body.

Halloween Also known as All Hallows Eve, it falls on 31 October, the eve of All Saints' Day, and is a special day at Hogwarts. In ancient Britain and Ireland the Celtic festival of Samhain was observed on that day, which marked the eve of the new year in both Celtic and Anglo-Saxon times. The last evening of October was 'old-year's night', carnival time for disembodied spirits. Huge bonfires were set on hilltops to scare them away. The souls of the dead were supposed to revisit their homes on this day, and ghosts, witches, hobgoblins and demons roamed about. This was the time to placate the supernatural powers that controlled nature.

Hex Once upon a time, to hex someone was to bewitch them. A hex could also be an evil spell, a symbol of bad luck or a witch. A hurling hex, for example, can cause broomsticks to buck. Today, of course, it stands for hexadecimal notation, a number system of base 16 used as a convenient way of representing the internal binary code of a computer.

Ice mice Sweets that make your teeth chatter and squeak. Muggles now know how to make them. A team led by Thomas Hofmann, deputy director of the German Research Centre for Food Chemistry in Garching, has developed a compound which has thirty-five times the cooling power of mint, the natural coolant most widely used in food. Unlike menthol, the cooling compound, one of a new class, was developed from chemicals in dark malt, the same material used to make beer and whisky. The compounds, known as cyclic alpha-keto enamines, are generally formed by reactions between sugars and amino acids when foods are heated. However, scientists have not explained all the magic of the mice. The exact mechanism by which these new cooling compounds exert their effect is unclear.

Immune system The range of weapons at the body's disposal to fight foreign forces, such as bacteria, viruses and fungi. Its role in dealing with charms and spells is unknown.

Knotgrass Also called allseed, a weedy plant whose small green flowers produce numerous seeds.

Lacewing Various insects with lacy wings that prey on aphids and other pests.

Lilac Gilderoy Lockhart's favourite colour.

Lionfish A spiny-rayed fish in which the spines are armed with venom glands.

Love potions and spells There are many examples in myth and legend. Take, for example, the Irish tale 'Diarmaid and Grainne', which dates back more than a millennium: Grainne, betrothed to be the chieftain's wife, falls for his nephew, Diarmaid, whom she enchants to elope. Today, one love potion is known as sildenafil, or Viagra. The drug blocks an enzyme called phosphodiesterase that itself destroys cyclic GMP, a molecule that makes blood vessels relax. Originally, the drug company Pfizer had hoped that this effect could boost blood flow to the heart, which is restricted in angina, by relaxing cardiac blood vessels. Instead, it was found to boost flood flow to the penis.

Mega Prefix denoting multiplication by a million.

Molecular biology The study of the molecular basis of life, including the biochemistry of molecules such as DNA and RNA.

Moody, Mad Eye Brain-scan studies have revealed that

moody people have a small area near the front of the brain – an inch or two behind the right eye (if you are right-handed) – that is probably working overtime. Called the ventromedial prefrontal cortex, it is thought to play a role in controlling heart rate, breathing, stomach acidity levels, sweating and similar autonomous functions that have a close connection with mood.

Mrs Skowers's All-Purpose Magical Mess Remover Probably relies on enzymes, similar to those found in powerful biological detergents. The enzyme, subtilisin, is a common laundry detergent additive that helps dissolve protein-containing stains, such as unicorn blood or gravy, by breaking the proteins into smaller, more soluble pieces.

Mutation A change in genes produced by a chance or deliberate change in the DNA that makes up the hereditary material of an organism.

Nanotechnology The building of devices on a molecular scale ('nanos' is Greek for dwarf). The ultimate in miniaturisation, it encompasses a range of methods, sensors and devices, anything that can be understood at the scale of nanometres, millionths of a millimetre. Spells that make objects appear out of thin air could, some speculate, be using silicon-chip-based machines, so tiny they can fit on the head of a pin, to grow an item, atom by atom, from the molecules in air and dust motes.

Neural networks Computers that can learn, loosely modelled on the vast interconnected networks of nerve cells (neurons) in the brain. The Sorting Hat may rely on one to interpret the patterns of activity in the wearer's brain.

Neuron The nerve cell that is the fundamental signalling unit of the nervous system.

Neurotransmitter A chemical that transmits signals between nerve cells.

Nucleic acid Complex organic acid made up of a long chain of units called nucleotides. The two types, known as DNA and RNA, form the basis of heredity.

Ordeal beans The Calabar bean is known as *esere* by the Efik people in Nigeria. Ingestion of the bean was used to substantiate or refute allegations, particularly of witchcraft. Survival was seen as evidence of innocence. Nearly all those who ate the Calabar bean died. Despite the toll of death and misery caused by the bean, it is a source of physostigmine, a drug used to prevent blindness in glaucoma patients.

Petrification Caused by a reflection of the murderous gaze of the basilisk. Scientists also know it as fibrodysplasia ossificans progressiva, a rare genetic disorder in which bone forms in muscles, tendons, ligaments and other connective tissues. Bridges of extra bone form across the joints in characteristic patterns, progressively restricting movement. It is estimated that FOP affects about 2,500 people worldwide, or approximately one in two million people.

Phase change A change in a material from one state into another, for example solid snowmen to liquid water, water to vapour, or solid to vapour.

Pheromone A chemical substance secreted by an animal that influences the behaviour of other animals. A few years ago, a University of Chicago researcher provided the first scientific proof of human pheromones: women's underarm odours can alter the timing of other women's reproductive cycles. Androstenol, occurring in human underarm sweat, has pheromonal qualities: females had 'increased social exchanges' with males after brief exposure to androstenol. Scientists have invested a great deal into finding out whether the pubescent Harry Potter will also be influenced by this kind of sexual chemistry. One pheromone, based on the underarm secretions of sexually active women, has already shown promise when it was used in a perfume in tests on men.

Phoenix The legendary and fabulous Arabian bird said to sing, set fire to itself and rise anew from the ashes every 500 years.

Photosynthesis Probably the most important chemical reaction on earth, because it enables plants to trap energy from the sun and convert it into a form that can sustain living cells. The key products of photosynthesis, which takes place in structures called chloroplasts, are glucose and oxygen. Eventually, that energy fuels creatures higher up the food chain, such as witches and wizards.

Poisons An impressive range are produced by bacteria, animals and plants in their attempts to kill or paralyse prey and predators. The female black widow is one of the most poisonous spiders, for example, thanks to the latrotoxin in its venom. The venom of the death adder contains toxins with a triple action: they stop blood from clotting, break down muscles and inactivate nerves. Some plankton, known as dinoflagellates, can contain maitotoxin, which can cause death at the level of one part per billion. One of the most feared bacterial toxins is botulinum toxin, probably the most toxic substance known to man. Produced by the bacterium *Clostridium botulinum*, the toxin affects the transmission of nerve impulses. Today, it is used to treat a vast array of medical conditions, most of which are the result of uncontrolled muscle contractions. Essentially, the toxin is used to turn off the signal that makes the muscles contract or stay contracted.

Polymer Enormously versatile molecules found throughout nature, as well as synthetic compounds with numerous commercial applications. They consist of long chains of atoms, and their physical versatility is derived from the variety in type, number and arrangement of those atoms. The chain is constructed from small repeating units, or monomers. Lignin, which makes the Whomping Willow woody, starch in potatoes and the cellulose in paper decorations are all examples of polymers. Perhaps the most important one is DNA, the genetic blueprint for living organisms.

Potion for dreamless sleep The ancient Greeks suspected that dreams were messages from the gods. Scientists have scant idea of why we sleep, and find dreaming an even bigger mystery. Theories range from the brain clearing its memory of junk to the liberation of suppressed subconscious urges. The potion must prevent REM sleep, a type of slumber which was discovered in 1953. During this phase of sleep the skeletal muscles are paralysed and nothing moves except the eyeballs, which swivel and rotate behind their closed lids. So instead of getting up and moving around as the brain instructs, the sleeper can dream. Pierre Maquet at the University of Liege, Belgium, has studied what happens during REM sleep and revealed activity in structures called the amygdalae, almond-shaped regions that play a role in the formation and consolidation of memories of emotional experiences. Among the hubbub within the sleeping brain, he also found reduced activity in an important region called the prefrontal cortex. To sleep, perchance to experience amygdalocortical activation and prefrontal deactivation?

Protein A class of large molecules found in living organisms, consisting of strings of amino acids folded into complex three-dimensional structures.

Put outer As used by Albus Dumbledore to extinguish streetlights. Light beams can be bent by gravity. There is now evidence that light can also become trapped. The suggestion came from studying the diffuse scattering of light that occurs in white paint, fog and clouds. For example, in snowflakes the rays of light are reflected from one crystal of ice to another, creating the white colour, before finding their way out again. But theorists predicted that when scattering became intense with only tiny distances between the reflections, the light waves would interfere with each other and become trapped in the material, a process called 'Anderson localisation'. Diederik Wiersma, of the European Laboratory for Non-Linear Spectroscopy in Florence, and colleagues were the first to witness Anderson localisation – semi-frozen light that bounces around in random loops – a few years ago, using gallium arsenide powder. However, he managed to store light for only one-thousandth of a millionth of a second, compared with the usual time of a millionth of a millionth of a second, so there is still some way to go before Muggles master the technology used by Dumbledore to snuff out streetlights.

Quantum theory The most revolutionary scientific theory of the last century, one that has to be taken into account in lasers, microelectronics and the flash of wands (see above). It was pioneered between 1900 and 1928 by European physicists, who realised that the previous 'classical' theories did not work when applied to subatomic particles. The reason is that only at this microscopic level does it matter that energy changes in abrupt and tiny jumps (quantum jumps). You can see when electrons make these jumps during firework displays or in candle flames; red light corresponds to a small quantum leap by an electron and blue to a relatively big leap.

Reaction In chemistry, the coming together of atoms or molecules with the result that a chemical change takes place, rearranging the way clouds of electrons buzz around colliding chemicals.

Receptor A receptor is a triggering device consisting of a protein in the membranes of our cells that allows drugs to work. Hormones and drugs are the keys that open a receptor's lock. When the chemical activates a given receptor, it will trigger a specific response. There are receptors in the tongue and nose and it is by stimulating them that Bertie Bott's beans achieve an astonishing repertoire of tastes.

Redcap In Muggle folklore, a Redcap was a spirit that haunted Scottish castles. Described as short, thickset old men, they would terrorise travellers after dark, sometimes butchering them and catching their blood in their caps.

Red hair Because the Weasleys have red hair, fair skin and freckles they could be a little bit Neanderthal. The origins of the 'ginger gene' date back around half a million years, according to a study by Rosalind Harding of the Institute of Molecular Medicine, Oxford University. Called the melanocortin 1 receptor, it is one of a group of genes that is responsible for black skin. Various mutations occurred in the receptor gene and, within the last 100,000 years, mutations arose in Europe that cause ginger hair and freckles. Two of these ginger mutations could be old enough to be Neanderthal. At the same time, this ancient ginger trait would not have thrived in Africa, where modern man emerged, because of its link with skin cancer. Modern man arrived in Europe 40,000 years ago, overlapping with the Neanderthals, who became extinct about 28,000 years ago, the last dying out in southern Spain and southwest France. The research suggests that the 'ginger gene' could have been passed down to modern Europeans from their Neanderthal ancestors. Thus the red hair, freckles and pale skin that characterise the Weasleys are – if scientists can confirm the ancient origins of the gene – the legacy of a long-dead species, known for being hairy and having prominent brows and receding foreheads.

Reductionism A doctrine according to which complex phenomena can be explained in terms of something simpler. In particular, atomistic reductionism contends that macroscopic phenomena, such as the heady aroma of warm butterbeer, can be explained in terms of the properties of atoms and molecules.

Relativity This theory deals with the concepts of space, time and matter and was developed mainly by Albert Einstein (1879–1955) as an extension of the ideas first set out by Isaac Newton. The effects predicted by relativity lie at the boundaries of our experience, in the domains of the super-small, the super-fast, the super-large and the super-massive. Special relativity, announced in 1905, starts from the premise that the speed of light and the laws of physics are the same – invariant – for observers moving at constant speeds relative to one another. General relativity, unveiled a decade later, broadens this idea so that the laws of physics should be the same for all observers, regardless of how they are moving relative to one another. Conceived after Einstein realised that a witch falling off a broomstick would not feel her own weight, general relativity replaced Newton's way of describing gravity, not as a force but the curvature of spacetime, a four-dimensional mix of space and time. Now more relevant than ever, given the interest in explaining how Portkeys and other wizardly forms of transportation work.

RNA (ribonucleic acid) The genetic material used to translate DNA into proteins. In some viruses, it can also be the principal genetic material.

Rowling, J.K. Witch who is able to cast enchanting literary spells.

Rubies (as big as glowing coals) Synthetic rubies were first made by the French chemist Auguste Verneuil in 1886, although the technique to make coal-size ones relatively easily was discovered by Polish scientist Jan Czochralski (1885–1953) in 1916. One evening he left aside a crucible with molten tin and returned to writing notes. At some moment, lost in his thoughts, instead of dipping his pen in the inkpot, he dipped it in the crucible and withdrew it quickly. He observed a thin thread of solidified metal hanging at the tip of the nib. Crystallisation began at the slot of the nib to form a single crystal.

In the Czochralski method developed from this insight, ingredient powders are melted in a platinum, iridium, graphite or ceramic crucible. A seed crystal is attached to one end of a rotating rod, the rod is lowered into the crucible until the seed just touches the melt, and then the rod is slowly withdrawn. The size of the boule (as the crystals are called) is determined primarily by the amount of material available in the crucible. Typical boules are 50 to 60 millimetres in diameter by as much as 500 millimetres long, said Keith Heikkinen of Saint-Gobain Crystals and Detectors in Washougal, Washington, where crystals are grown for lasers. Compared to a laser-grade boule there is about fifty times more chromium added to jewellery material and the crystalline quality is far less rigorous and the boule more massive: around 12 centimetres across, 40 centimetres long and 10 kilograms – the equivalent of 50,000 carats.

Runes Characters of an ancient Germanic alphabet that were thought to have magical significance, though the term is usually used to describe any obscure piece of writing with mysterious symbols, such as theoretical physics papers.

Salamander A fire-dwelling lizard, according to Fred and George Weasley. Muggles know of many species, such as the American hellbender (*Cryptobranchus alleganiensis*), which has a strong bite and exudes a noxious slime, or the European fire salamander (*Salamandra salamandra*), which produces toxic secretions from glands behind its eyes.

Scar When Harry was attacked by Voldemort on Halloween in 1981, he was branded with a lightning-bolt scar by an evil curse. Like other scars, the zigzag on Harry's forehead is best defined as a macroscopic mark caused by a previous injury which has a structural basis in a disorganised pattern and perhaps an excessive amount of extracellular matrix molecules (mostly the proteins collagen and fibronectin), according to Mark Ferguson of the University of Manchester. The abnormal organisation is due to the abnormal migration of fibroblasts into the wound site, and is controlled by among other things a protein called TGF β. Disorganised fibres cause abnormal forces which of course give rise to bumps and valleys recognised within the macroscopic scar. Unfortunately for Harry, he was scarred by Voldemort before new advances were made in the science of scarring, after studies into why it does not occur in tissue damaged before birth. Ferguson's group was the first to discover the cellular and molecular mechanisms responsible for scar-free healing in the embryo, and the team has identified a number of the major cellular and molecular targets for drugs to prevent scarring in the adult. On the basis of this work, Ferguson has founded a university spin-off company, Renovo. However, it is still something of a puzzle how Dumbledore ended up with a scar above his left knee that is a perfect map of the London Underground.

Scarab beetles The best-known example is the dung beetle, revered in ancient Egypt, where in hieroglyphs it was the symbol of the solar deity. The beetle rolls pieces of dung into balls in which it lays its eggs. The ball rolling was thought to symbolise the movement of the sun.

Second Law of Thermodynamics The writer C.P. Snow said that this law should be part of the intellectual complement of any well-educated person. Put baldly, the law states that glasses of butterbeer do not spontaneously heat up or puddles of water cool down to form snowmen.

Shooting stars Seen after Voldemort's first tussle with Harry Potter, shooting stars occur when earth travels through comet dust and the debris burns up in the atmosphere. One well-known example is the Leonid shower, which occurs when earth passes close to the orbit of the comet Tempel-Tuttle and the debris left in the comet's path. Some of these dust streams actually broke away from the comet long ago. Meteors visible in 2000 dated to 1932, 1866 and 1733.

Sphinx The sphinx of Greek mythology had the head and breasts of a woman, the body of a dog or lion, the wings of a bird, a tail, and lion's paws. She lived at the gates of Thebes and, as in *Harry Potter and the Goblet of Fire*, set riddles to passers-by. She devoured those who were unable to solve them. The locals were told by the oracles that the Sphinx would kill herself if the following was solved: What goes on four feet, on two feet, and three, but the more feet it goes on the weaker it be? It was solved by Oedipus. The answer is: a man (who as a baby crawls on all fours, in manhood walks on two feet, but in old age needs the support of a staff).

Thermodynamics The science of heat and work. In a gas, for example, it can relate properties such as heat, temperature and pressure by taking into account how the atoms and molecules of the gas behave like Bludgers.

Tickling charm The psychologist Wallace Chafe suggests that laughter serves to incapacitate us, acting as a disabling device that allows you to relax when you realise that a threat is not genuine. Vilayanur Ramachandran of the University of California, San Diego, says that this idea, though intriguing, does not explain our need to generate 'rhythmic, loud, explosive sounds'. In other words, it may help us relax but why tell others about it? We chortle to announce that a perceived threat or anomaly is trivial. 'You approach a child, hand out menacingly . . . But no, your fingers make light, intermittent contact with her belly,' he writes. 'As a result of this tickle, the child laughs, as if to inform other children, "He doesn't mean harm".' Unfortunately, it is not possible to tickle yourself if you lack a companion. Something in the brain, given the humourless label of a 'corollary discharge', anticipates the effects of our actions and dulls the sensation.

Tubeworms At Hogwarts, these sometimes have to be scraped off desks. On earth, they are some of the planet's strangest life-forms. They are found at seeps and vents, regions on the ocean floor where fluids rising from the earth's crust break through and mix with seawater. The vents have a wide range of shifting environments including 'black-smoker' chimneys belching out material as hot as 400 degrees Celsius, whereas seeps of hydrocarbons are at cold temperatures typical of seawater. Worms around the latter can take up to 250 years to grow 6 feet long, while the former grow well over 6 feet in just one year. There are various types. For example, one vent tubeworm (*Riftia pachyptila*) looks like a giant lipstick and can grow to 9 feet tall, while the hairy, 5-inch Pompeii worm (*Alvinella pompejana*) currently holds the record as the 'hottest' animal on earth.

Veela The most beautiful women that Harry had ever seen (unless they are provoked, when their faces elongate to form bird beaks and long scaly wings begin to sprout). Scientists now understand what happens in the brain when one makes eye contact with an attractive person such as a Veela – it boosts activity in the brain's ventral striatum, which is linked with the expectation of a reward, according to Knut Kampe of University College London, and colleagues.

Virus One of the smallest infectious agents, consisting of a piece of genetic code wrapped in protein, measuring between 15 and 300 nanometres across (one nanometre is one thousand-millionth of a metre). They are responsible for a huge range of diseases, such as influenza. It is debatable whether viruses are living, since they have to hijack the molecular machinery of our cells to reproduce (they do this by 'reprogramming' our cells with their genetic code, turning them into virus factories). That is why the common cold is so tricky to treat: it is difficult to combat them with drugs without also harming the cells they parasitise, though safe effective drugs do exist for some viral infections. One of them is Madam Pomfrey's Pepperup Potion. She dispensed quite a lot of this as flu season struck Hogwarts. It leaves the patient's hair smoking.

Waterproof fire A speciality of Hermione Granger. Muggles demonstrated this is indeed possible in the year 2000 when marine biologist Wendy Craig Duncan surfaced with the Olympic flame still burning after swimming 100 metres underwater along Agincourt Reef, part of Queensland's Great Barrier Reef. The flare was developed to fit into the boomerang-shaped Sydney 2000 Games Olympic Relay Torch by Pains Wessex Australia's pyrotechnic team in Melbourne. The flare reached 2,000 degrees Celsius under water. Sufficient oxygen had to be generated to allow its main light-producing magnesium fuel to burn smoothly in a gas envelope in order to provide a bright underwater flame, while giving off enough gas to prevent seawater from entering the flare's tube and extinguishing the flame.

If Newton had not, as Wordsworth put it, voyaged through strange seas of thought alone, someone else would have. If Marie Curie had not lived, we still would have discovered the radioactive elements polonium and radium. But if J.K. Rowling had not been born, we would never have known about Harry Potter. That is why Master Potter means so much to all of us. Science may be special but Harry, as a work of art, is more so. Harry Potter is unique.

Index